Power Systems

For further volumes:
http://www.springer.com/series/4622

Tadasu Takuma • Boonchai Techaumnat

Electric Fields in Composite Dielectrics and their Applications

Springer

Dr. Tadasu Takuma
Atlas Tower 710, Noborito 2130-2
Tama-ku, Kawasaki-City
Kanagawa-Pre.
214-0014 Japan
tctaku@sb.dcns.ne.jp

Dr. Boonchai Techaumnat
Chulalongkorn University
Fac. Engineering
Dept. Electrical Engineering
Phyathai Road
10330 Bangkok
Thailand
boonchai.t@chula.ac.th

ISSN 1612-1287 e-ISSN 1860-4676
ISBN 978-90-481-9391-2 e-ISBN 978-90-481-9392-9
DOI 10.1007/978-90-481-9392-9
Springer Dordrecht Heidelberg London New York

Library of Congress Control Number: 2010933730

Cover design: SPI Publisher Services

Printed on acid-free paper

Springer is part of Springer Science+Business Media (www.springer.com)

Acknowledgements

We wish to express our appreciation to our colleagues who are the coauthors of our papers cited in this book. Among them, we would particularly like to thank Prof. Teruya Kouno, Assoc. Prof. Shoji Hamada, the late Dr. Tadashi Kawamoto, Dr. Hideo Fujinami, Dr. Masami Kadonaga, and Prof. Bundhit Eua-arporn for their cooperative work and suggestions. We also wish to thank Mr. David Smallbones for his careful editing of our manuscripts.

Contents

Chapter 1
Basic Properties of Electric Fields in Composite Dielectrics

1.1 Background

An accurate quantitative picture of electric field distribution is essential in many electricity-related areas and applications. Some typical examples are the analysis of discharge phenomena and their application, insulation designs for high-voltage power equipment, designs for electrostatic devices and devices used for high field emission or electron beam generation, and assessing the various environmental issues associated with electric fields.

The computerization of numerical field calculation techniques with high-capacity computers now enables us to evaluate electric field distributions quantitatively and precisely for any configuration of elements, so long as only one dielectric medium and conductors are involved. In these monodielectric cases, the actual field distribution more or less matches our expectations.

In composite dielectric media composed of multiple dielectrics, the situation is quite different. Anomalous or unexpected behavior may appear, including significant enhancement of the electric field at a point where a solid dielectric is in contact with a conductor (an electrode) or another solid dielectric. Accurate knowledge of the field in the contact region is important, since it is directly related to a wide variety of practical applications. Some typical examples are:

- Design of high-voltage equipment with solid insulating supports to avoid the initiation of discharge due to field enhancement
- Utilization of the electrostatic force on dielectric particles, as in electrophotography, and electrorheological (ER) fluids
- Applications of field emission, such as electron beam devices in a vacuum

One specific phenomenon or behavior in the contact field is sometimes called the triple-junction (or triple-joint) effect because the contact point is where three media meet: a conductor, a solid dielectric, and a gaseous dielectric (or a liquid or a vacuum). While such points of contact always appear in a system composed of a gas, liquid or vacuum (usually acting as an insulator); an energized conductor; and

T. Takuma and B. Techaumnat, *Electric Fields in Composite Dielectrics and their Applications*, Power Systems, DOI 10.1007/978-90-481-9392-9_1,

a solid insulator (inevitably required to support the conductor), it is only in recent years that the field behavior has been considered important enough to attract a great deal of attention. This is principally for the following reasons:

1. The development of compressed SF_6 gas and vacuum insulation systems in the electric power sector.

Both insulation systems consist of dielectric media with very high resistivity under perfectly dry conditions. In contrast, in the open-air insulation systems of conventional power equipment, the field behavior is less important due to the effect of the inevitably present surface conductivity and possible partial discharge, which may reduce or nullify the field enhancement at the contact points.

2. The development of electrostatic applications such as electrophotography and ER fluids.

The design and analysis of these applications require in-depth knowledge of the forces governing adhesion (and also detachment) and how they relate to the field at contact points.

3. Progress in numerical field calculation.

Although some information has been available about field behavior at contact points for a few basic configurations, the recent development of numerical field calculation techniques has revealed interesting features and added quantitative information about a wide variety of more complex situations.

1.2 Fundamentals of Composite Dielectric Fields

1.2.1 Governing Equations

The basic equation for electric field **E** is derived from one of Maxwell's equations:

$$\mathrm{rot}\,\mathbf{E} = -\frac{\partial \mathbf{B}}{\partial t} \tag{1.1}$$

In quasi-electrostatic cases with the magnetic field **B** being zero or constant on the right-hand side of Eq. 1.1, the electric field **E** has a scalar potential ϕ, i.e.,

$$\mathbf{E} = -\mathrm{grad}\,\phi \tag{1.2}$$

Combining Eq. 1.2 with another of Maxwell's equations relating electric flux density **D** and charge density q,

$$\mathrm{div}\,\mathbf{D} = q \tag{1.3}$$

and a constituent equation for the materials,

$$\mathbf{D} = \varepsilon\varepsilon_0\mathbf{E} \tag{1.4}$$

leads to the following equation for ϕ:

$$\text{div}\,(\varepsilon\varepsilon_0\text{grad}\ \phi) = -q \tag{1.5}$$

In this equation, ε is the dielectric constant (relative permittivity) of the material or medium, ε_0 is the electric constant, and q is the space charge density. It should be noted that ε is dimensionless while ε_0 is 8.854×10^{-12} F/m.

Throughout this book, ε is considered to be constant inside each medium, although in practice, the dielectric constant ε of solid and liquid materials changes somewhat depending on the applied electric field. Under the approximation of constant ε and in the absence of space charge, the governing equation for ϕ becomes Laplace's equation inside each medium:

$$\text{div (grad } \phi) = \Delta\phi = 0. \tag{1.6}$$

The main task in most electric field calculations is to solve the electric potential ϕ which satisfies Eq. 1.6 and the corresponding boundary conditions.

1.2.2 Boundary Conditions

The boundary condition on conductor surfaces is a constant electric potential, $\phi = \phi_{0,}$ the so-called Dirichlet condition. At the interface of two dielectrics ε_A and ε_B, however, two conditions hold. One is the continuity of ϕ, or equivalently the continuity of tangential field strength E_t,

$$\phi_A = \phi_B \tag{1.7}$$

or

$$E_{tA} = E_{tB} \tag{1.8}$$

where subscripts A and B indicate the corresponding media or materials. The other condition is the continuity of the normal component D_n of the electric flux density,

$$D_{nA} = D_{nB}. \tag{1.9}$$

From Eq. 1.4, this can be rewritten as

$$\varepsilon_A E_{nA} = \varepsilon_B E_{nB}. \tag{1.10}$$

These equations, together with the constant potential condition on conductor surfaces, complete the boundary conditions for composite dielectric configurations. However, it should be noted that these boundary conditions at dielectric interfaces are only macroscopic expressions and should not be generalized to such minute surface structures as those involving molecules or atoms.

1.3 Effect of Conduction

Any real dielectric exhibits a certain degree of conductivity, producing what are usually called capacitive-resistive or mixed fields. Apart from electrical discharges, two kinds of conduction are possible: volume conduction and surface conduction, which are characterized by volume and surface conductivities, σ and σ_s, respectively. Surface conduction is in fact volume conduction in a very thin surface layer of a solid material, which can conveniently be treated as being of zero depth. Such surface conduction is usually brought about by adsorbed humidity or other contamination. Although both volume and surface conductivities may change locally depending on the conditions, they are treated as constant in each medium throughout in this book.

1.3.1 Basic Equations

Before explaining capacitive-resistive (mixed) fields, we will briefly mention the treatment with complex variables (phasor notation) for steady ac fields. When all the materials are perfect dielectrics without conductivity and only one kind of voltage waveform (source) is present, field calculation can be performed as for a dc field, as long as the applied voltage does not change very fast compared to the traveling time of the field. In slowly changing field conditions, the instantaneous potential and field strength are always similar in any part of the region, i.e., they are merely proportional to the applied voltage at that instant. When several voltage sources exist, the field must be computed not as a dc field but by superposing the instantaneous effects of all the sources. If all the waveforms of the sources are sinusoidal with the same angular frequency ω, we can conveniently express the instantaneous field in the system directly with complex field values.

We can calculate mixed fields by taking into consideration the true charge induced by volume or surface conduction. In a region with dielectric constant ε and volume conductivity σ, electric field $\mathbf{E}$ is formulated from Eqs. 1.3 and 1.4 as follows:

$$\operatorname{div}(\varepsilon\varepsilon_0\mathbf{E}) = q. \tag{1.11}$$

On the other hand,

$$\operatorname{div}\mathbf{J} = \operatorname{div}(\sigma\mathbf{E}) = -\frac{\partial q}{\partial t} \tag{1.12}$$

where **J** is the current density and q is the charge density in the space. These equations lead to

$$\operatorname{div}\left(\sigma\mathbf{E} + \varepsilon\varepsilon_0\frac{\partial\mathbf{E}}{\partial t}\right) = 0. \tag{1.13}$$

In the steady state for ac fields of angular frequency ω ($= 2\pi f$, where f is the corresponding frequency), the phasor notation (complex number expression) can be used to represent the electric field distributions. Thus, Eq. 1.13 becomes

$$\operatorname{div}\left\{(\sigma + \mathrm{j}\omega\varepsilon\varepsilon_0)\dot{\mathbf{E}}\right\} = 0 \tag{1.14}$$

where $\mathrm{j} = \sqrt{-1}$ and the overdot indicates a complex expression (phasor notation) of the variable for ac fields. The field, including the effect of volume conduction, can be expressed simply by substituting for dielectric constant ε a complex expression, such as

$$\dot{\varepsilon} = \varepsilon + \frac{\sigma}{\mathrm{j}\omega\varepsilon_0}. \tag{1.15}$$

In cases without conduction, $\sigma = 0$ (infinitely high resistivity) and $\dot{\varepsilon}$ simply reduces to ε. However, Eq. 1.15 indicates that the effect of volume conduction is predominant for a dc field or ac fields where $\sigma \gg \omega\varepsilon\varepsilon_0$.

If dielectric constant ε and volume conductivity σ are both constant in a medium or a part of the region, the governing equation, Eq. 1.14, once again becomes Laplace's equation as given in Eq. 1.6 for the complex electric field. In this condition, an induced charge does not exist inside each medium but only at its boundaries (i.e., on conductor surfaces and dielectric interfaces).

1.3.2 Boundary Conditions

The boundary conditions given in Eqs. 1.7–1.10 must be somewhat modified at the dielectric interfaces in mixed fields. The continuity of electric potential ϕ, Eq. 1.7, or the continuity of tangential field strength E_t, Eq. 1.8, still applies. However, the continuity of the normal component of the electric flux density, Eq. 1.9 or 1.10, changes as follows [1].

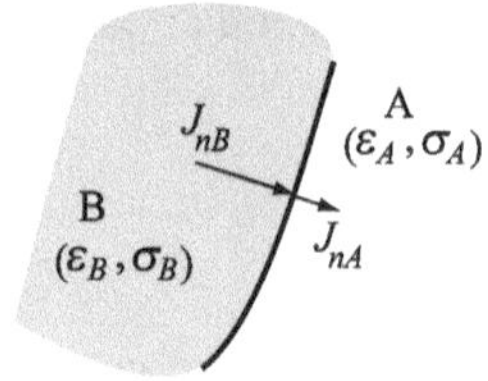

Fig. 1.1 Interface of dielectrics with volume conduction

For systems with two dielectrics with volume conductivity, i.e., dielectric A (dielectric constant ε_A, volume conductivity σ_A) and dielectric B (ε_B, σ_B), as shown in Fig. 1.1, the continuity condition of the normal component of the electric flux density at the interface is expressed as

$$D_{nA} - D_{nB} = q_s. \tag{1.16}$$

In this equation, subscript n means a component normal to the interface, and q_s is the surface charge density, which is given by

$$q_s = \int (J_{nB} - J_{nA}) dt. \tag{1.17}$$

Since current density **J** is equal to σ**E**,

$$q_s = \int (\sigma_B E_{nB} - \sigma_A E_{nA}) dt. \tag{1.18}$$

We can rewrite the integral with 1/jω for ac fields with an angular frequency ω in the phasor notation, thus

$$\dot{q}_s = \frac{1}{\mathrm{j}\omega} \left(\sigma_B \dot{E}_{nB} - \sigma_A \dot{E}_{nA} \right). \tag{1.19}$$

This equation, when combined with Eq. 1.16, gives the final boundary condition as

$$\left(\varepsilon_A + \frac{\sigma_A}{\mathrm{j}\omega\varepsilon_0} \right) \dot{E}_{nA} = \left(\varepsilon_B + \frac{\sigma_B}{\mathrm{j}\omega\varepsilon_0} \right) \dot{E}_{nB}. \tag{1.20}$$

Equation 1.20 is the same as that for the interfaces of usual dielectrics without conductivity with ε replaced with the complex dielectric constant given in Eq. 1.15.

When surface conduction exists, a more general conservation of charge is expressed as [2]

$$\dot{\varepsilon}_B \dot{E}_{nB} - \dot{\varepsilon}_A \dot{E}_{nA} = \frac{1}{\mathrm{j}\omega\varepsilon_0} \nabla_s \cdot \dot{\mathbf{J}}_s \tag{1.21}$$

The effect of surface conduction along a dielectric interface is included in the right-hand side of Eq. 1.21, where $\dot{\mathbf{J}}_s = \sigma_s \dot{\mathbf{E}}_t$ is the surface current density at the interface and $\nabla_s \cdot \dot{\mathbf{J}}_s$ is the tangential divergence of $\dot{\mathbf{J}}_s$. The normal is taken in the direction from medium B to medium A.

1.3.3 Classification Based on the Effect of Volume Conduction

When characteristic variables $\omega\varepsilon\varepsilon_0$ and σ are constant in each medium, Eq. 1.14 is classified into the following three categories according to the defining parameter $\omega\varepsilon\varepsilon_0/\sigma$:

(a) $\omega\varepsilon\varepsilon_0/\sigma \gg 1$: high frequency or capacitive fields

$$\mathrm{div}\left(\varepsilon\dot{\mathbf{E}}\right) = 0 \tag{1.22}$$

This equation corresponds to usual electric fields where an ac voltage is applied.

(b) $\omega\varepsilon\varepsilon_0/\sigma \ll 1$: dc (low frequency) or resistive fields

$$\mathrm{div}\left(\sigma\dot{\mathbf{E}}\right) = 0 \tag{1.23}$$

This equation is the same as Eq. 1.22 where ε is replaced with σ, and the corresponding field is often called a steady current field.

(c) General cases or mixed fields

The field equation, Eq. 1.14, is equal to Eq. 1.22 where ε is replaced with the complex dielectric constant given in Eq. 1.15. If both ε and σ are constant everywhere, Eq. 1.14 again leads to $\mathrm{div}\,\dot{\mathbf{E}} = 0$, i.e., Laplace's equation.

1.4 Outline of Field Behavior near a Contact Point

1.4.1 Contact Angle

For simplicity, in most parts of this book, we employ a composite dielectric configuration consisting of two dielectrics with dielectric constants (relative permittivities) ε_A and ε_B, respectively. Only in Chapter 4 and parts of Chapters 5–7 do we deal with systems with three dielectrics.

In fundamental cases where the interface of two dielectrics ε_A and ε_B meets a conductor surface (or an electrode surface at a fixed potential), the most important factor that characterizes the field behavior near the point of contact is the contact angle formed by the interface with the conductor surface [3]. Figure 1.2 illustrates three categories of contact angle α that are explained below. Some simple, practical examples of composite dielectrics are shown in Fig. 1.3.

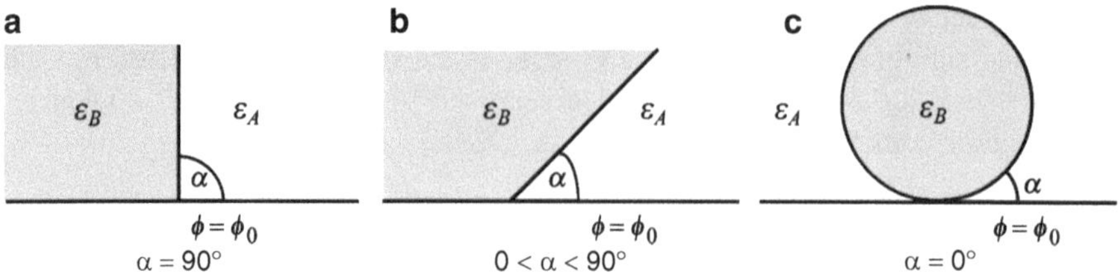

Fig. 1.2 Classification of contact angle

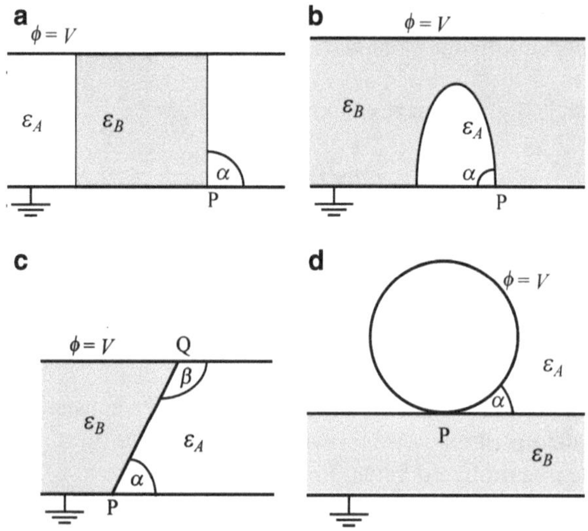

Fig. 1.3 Examples of simple contact geometries in composite dielectric configurations

Field behavior is basically summarized according to α as follows:

(1) $\alpha = 90°$

No field singularity occurs in this case. Furthermore, if the dielectric interface is everywhere coincident with (parallel to the direction of) electric lines of force in the corresponding configuration without the interface, i.e., in the monodielectric case of either ε_A or ε_B, the field distribution remains unaffected, being identical to the original situation without the interface. Figure 1.3a shows one such simple example. Needless to say, the contact angle of the corresponding interface must be equal to 90° because electric lines of force always make a right angle with a conductor surface in any monodielectric case.

When the dielectric interface does not everywhere coincide with the original electric lines of force, its presence alters the field distribution to a greater or lesser extent, even for $\alpha = 90°$. We will deal with such typical examples for the case $\alpha = 90°$ below in Section 1.4.2.

(2) $0 < \alpha < 90°$

The dielectric interface meets the conductor surface as a straight line in the cross-sectional view, making an angle $\alpha < 90°$ (or $90° < \alpha < 180°$) with respect

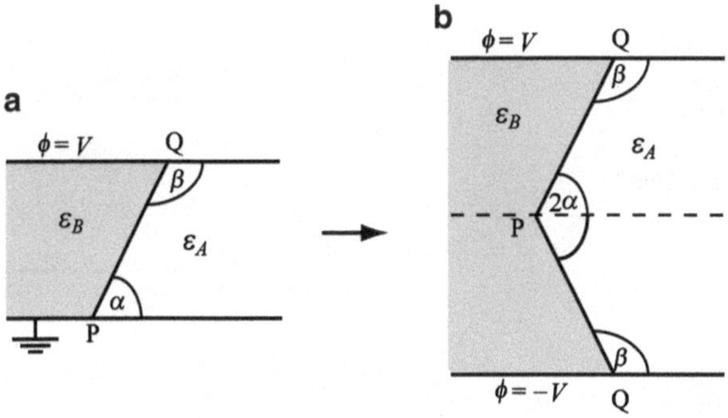

Fig. 1.4 Configuration with a mirror image of Fig. 1.3c

to the plane of the conductor surface. In this configuration, the field strength theoretically exhibits a singularity (an infinitely high field) or is zero at the contact point, and the field behavior is discussed in Chapter 2. One variation of this case is the configuration where the interfaces of three dielectrics ε_A, ε_B, and ε_C meet at a common point, as encountered in a solid dielectric supporting another solid. This case is discussed in Chapter 4.

(3) $\alpha = 0°$

In this case, either the dielectric interface or the conductor surface (or both) is curved, thus having a common tangential line of contact. This produces a contact angle $\alpha = 0°$. This configuration does not cause an infinitely high field but involves considerable field enhancement at the point of contact, and is discussed in Chapter 3.

Two important points should be noted here. First, the contact conditions as exemplified in Figs. 1.2 and 1.3 may represent both two-dimensional (2D) and axisymmetric (AS) systems. In Fig. 1.3d, for example, the upper conductor (electrode) may be a sphere (an AS case) or a cylinder (a 2D case). Second, when these configurations have a grounded plane, the identical field distribution can be realized by adding the mirror image (plane-symmetrical) configuration below the grounded plane. One example is illustrated in Fig. 1.4, in which the field is the same in configurations (a) and (b), resulting in a field singularity (an infinitely high field) at point P on the equipotential surface (dashed line) in Fig. 1.4b. Configuration (b) contains no contact point in the middle, and so there is no triple junction there: it is composed of only two dielectrics, lacking a conductor as the third medium.

1.4.2 *Typical Examples for Contact Angle* $\alpha = 90°$

1.4.2.1 Axisymmetric (AS) Cases

For the purpose of comparison with the field behaviors described in the following chapters, we briefly explain here the field in the typical case of $\alpha = 90°$.

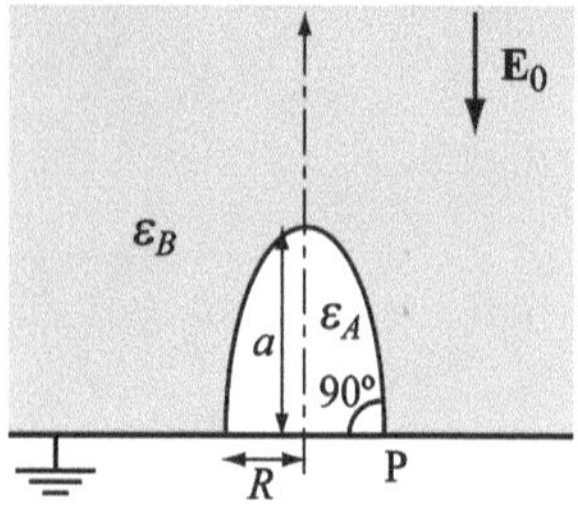

Fig. 1.5 Semi-ellipsoid of revolution with dielectric constant ε_A in a semi-infinite region with dielectric constant ε_B on a plane conductor under uniform field E_0

The configuration in question is shown in Fig. 1.5; it consists of a semi-ellipsoid of revolution (or semi-spheroid) as an example of an AS case with dielectric constant ε_A existing in another dielectric ε_B on a grounded plane under a uniform field E_0. This case also has practical importance as a model configuration, e.g., of a dielectric void. As explained above in the context of Fig. 1.4, the field distribution of Fig. 1.5 is identical to that for an ellipsoidal void with dielectric constant ε_A existing isolated in an infinite region with dielectric constant ε_B under a uniform field.

The field distribution can be solved analytically using the ellipsoidal coordinates. The electric field is uniform everywhere inside the semi-ellipsoid of Fig. 1.5 (or ellipsoid in the infinite region), and can be expressed for the AS case as

$$E = \frac{\varepsilon_B E_0}{\varepsilon_B + (\varepsilon_A - \varepsilon_B)C}. \tag{1.24}$$

In this expression,

$$C = \left(\frac{1}{e^2} - 1\right)\left(\frac{1}{2e}\ln\frac{1+e}{1-e} - 1\right), \quad e = \sqrt{1 - R^2/a^2} \ \text{ for } a > R, \text{ and}$$

$$C = \frac{1}{e^2} - \frac{\sqrt{1-e^2}}{e^3}\arcsin e, \quad e = \sqrt{1 - a^2/R^2} \quad \text{for } a < R.$$

For the special case of a hemisphere ($a = R$),

$$E = \frac{3\varepsilon_B E_0}{\varepsilon_A + 2\varepsilon_B}. \tag{1.25}$$

Figure 1.6 represents normalized field strength E/E_0 inside the (semi-) ellipsoid for ε_s ($= \varepsilon_B/\varepsilon_A$) ranging from 2 to 10 and a/R ranging from 10^{-3} to 10. These results confirm the explanation found in most textbooks on electromagnetism that the field inside a spheroid (or gaseous void in practice) is nearly equal to E_0 for a thin void ($a \gg R$) oriented in the direction of the original (external) field, while it is multiplied by ε_s for a flat void ($a \ll R$) that is perpendicular to the field. Needless to say, these relations can be derived in an analytical way by considering the infinitely large or small limit of a/R in Eq. 1.24.

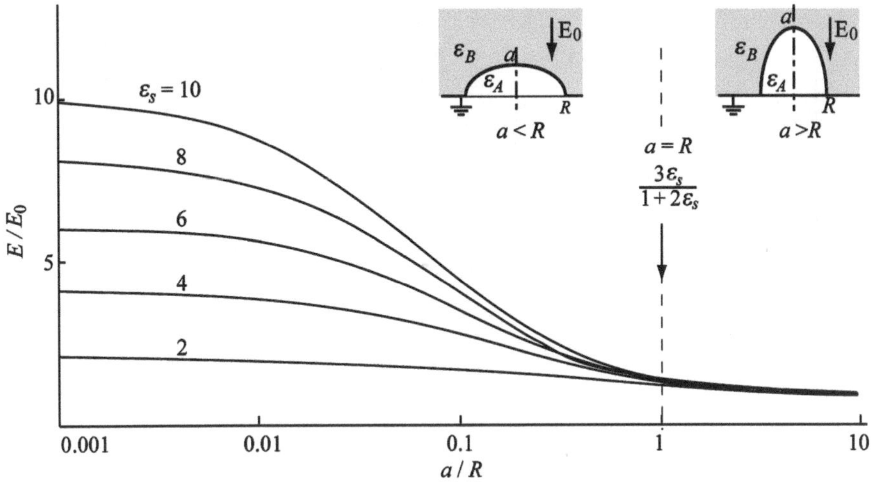

Fig. 1.6 Normalized field strength inside ε_A in Fig. 1.5 for semi-ellipsoids (AS cases) [3]. © 1991 IEEE

1.4.2.2 Two-dimensional (2D) Cases

Field behavior similar to that shown in Fig. 1.6 is also found in 2D cases consisting of a semi-elliptic column instead of the semi-ellipsoid in Fig. 1.5. The electric field in this case is also uniform everywhere inside the semi-elliptic column, and is given as

$$E = \frac{(R+a)\varepsilon_B E_0}{a\varepsilon_A + R\varepsilon_B}. \tag{1.26}$$

For the special case of a semi-column ($a = R$),

$$E = \frac{2\,\varepsilon_B E_0}{\varepsilon_A + \varepsilon_B} \tag{1.27}$$

Figure 1.7 represents normalized field strength E/E_0 inside the (semi-)elliptic column for conditions corresponding to those in Fig. 1.6.

1.5 Outline of the Chapters

This book deals with electric fields (also electrostatic force in some cases) in composite dielectrics, with the emphasis on various contact conditions between solid dielectric interfaces and the surface of a conductor (an electrode) or another dielectric. The content is divided into the following four parts.

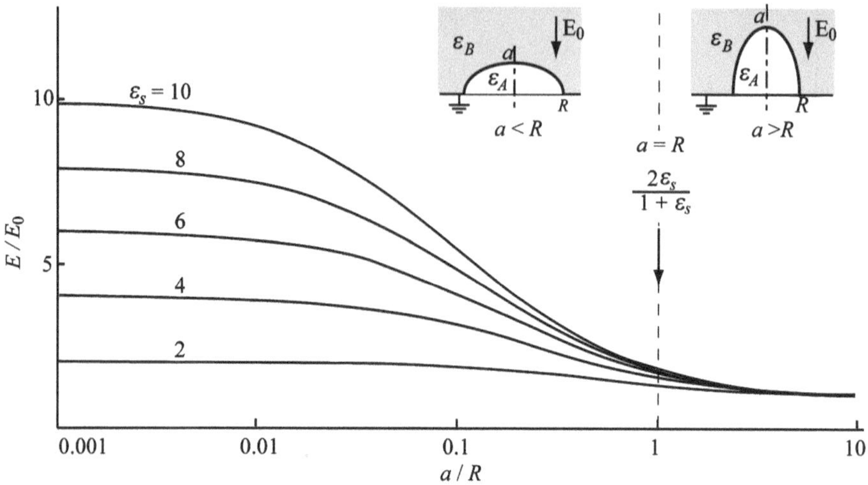

Fig. 1.7 Normalized field strength inside ε_A in Fig. 1.5 for semi-elliptic columns (2D cases)

- Part A: Chapter 1 covers the relevant fundamental subjects.
- Part B: Chapters 2–4 explain the basic properties of electric fields for three kinds of contact conditions.
- Part C: Chapters 5–7 deal with the applications of field analysis in three areas.
- Part D: Chapters 8 and 9 explain the calculation methods used to analyze electric fields in composite dielectrics.

Chapter 1 deals with such subjects as the background of the book, the governing equations with the boundary conditions, the effect of conduction, and the classifications of field behavior according to the contact angle. It additionally explains the behavior for a contact angle of 90°.

Chapter 2 explains the electric fields for a contact angle between 0° and 90°, i.e., for cases in which the cross-sectionally straight interface of two solid dielectrics meets a conductor surface. The electric field is proportional to l^m near the point of contact, where l is the distance from the contact point; this means that the field strength theoretically becomes infinitely high or zero at the point of contact. Values of m are derived for various configurations by analytical and numerical field calculation methods. This chapter also examines the field when an interface edge is rounded to realize right-angled contact as well as the effect of volume and surface conduction.

Chapter 3 deals with the electric fields for smooth contact with a zero contact angle, i.e., when a conductor and a solid dielectric have a common tangent at the point of contact, forming no sharp edge. This contact condition does not result in an infinitely high field strength but does give significant field enhancement at the point of contact. The electric field distribution is explained for simple exemplary configurations, e.g., arrangements of a stressed conducting sphere or cylinder lying on a dielectric solid and an uncharged conducting sphere under a uniform field.

To summarize, this chapter presents the approximate expressions of the contact point field for configurations with a zero contact angle. Also examined are the effects of volume and surface conduction for some basic configurations.

Chapter 4 deals with the electric fields for configurations in which the straight interfaces of three dielectrics meet at a point. As was done for arrangements with two dielectrics in Chapter 2, the electric field is expressed as l^m near the point of contact, where l is the distance from the contact point. This chapter presents equations for determining exponent m, that are much more complex than those for the two-dielectric case. Also described are the behaviors for a solid dielectric supporting another dielectric and for smooth contact with a rounded interface edge.

Chapter 5 discusses some related topics in the field of high-voltage equipment. Composite dielectric interfaces frequently result in weak points in insulation systems, especially those due to field enhancement at points of contact. This chapter first describes field enhancement and relevant preventive measures for solid insulating supports, so-called spacers, in gas-insulated coaxial structures. Next, the techniques of optimizing a solid dielectric shape are briefly mentioned. The field distribution is studied for two typical arrangements often encountered in practice: a flange structure and a solid dielectric rod support. Extending the analysis given in Chapter 4, the field for a dielectric solid supporting another solid is investigated. This chapter also discusses some applications involving high field emission in a vacuum.

Chapter 6 reports the calculated results of the electric field and force in systems of multiple (spherical) particles. These results are generally applicable to so-called pearl-chain-forming force problems, but their direct application focuses on electrorheological (ER) fluids, i.e., liquid media that exhibit a significant change in rheological properties when subjected to an electric field. After explaining the approximate calculation with equivalent dipoles, the correct numerical results for multiple particles are given, including a particle chain tilted to the field direction. A chain of two particles, which is isolated in space, in contact with a plane electrode, or existing between parallel plane electrodes, is studied in detail as its behavior approximates that of individual particles in a chain of many particles. The chapter also explains the results for nonhomogeneous particles, i.e., particles coated with a surface film.

Chapter 7 describes the electric field and force on a charged particle (e.g., toner in electrophotography), particularly, when the charge distribution is nonuniform. The practical background, the meaning of the calculation, and then the particular application of the re-expansion method for analyzing the field in this chapter are explained. The principal point of the chapter concerns the force behavior of a particle on a grounded plane, focusing on adhesion without an applied field and detachment by an external field. The calculation is carried out mainly for two non-uniform surface charge distributions: the bottom-pole distribution and the two-pole distribution. Finally, the force characteristics of a charged dielectric particle on a dielectric barrier are explained.

Chapter 8 deals with the analytical methods used to calculate the electric field in composite dielectrics, in particular, analysis of the field near points of contact. Three methods are explained: the variable-separation method for cross-sectionally straight dielectric interfaces, the iterative image charge method for a conducting sphere or cylinder on or above a solid dielectric plane, and the re-expansion method for a chain of multiple spherical particles.

Chapter 9 deals with the numerical methods used to calculate the electric field in composite dielectrics. First, the basic features of various numerical methods for determining electric fields are reviewed, focusing on the differences when they are applied to calculations for composite dielectrics. Three methods that are mainly used to analyze the field in composite dielectrics are described: the charge simulation method (CSM), the surface charge method (SCM), and the boundary element method (BEM). For composite dielectric configurations, the principles of the three methods are explained, some improvement techniques are discussed, and a few exemplary subdivision states in numerical calculations are given for composite dielectric configurations.

Chapters 2–9 contain a separate section entitled Introduction at the beginning. Most of these Introductions give a brief historical overview of the subjects discussed in the rest of the chapter, while the others give an overview of the content of the chapter.

References

1. Takuma, T., Kawamoto, T., Fujinami, H.: Charge simulation method with complex fictitious charges for calculating capacitive-resistive fields. IEEE Trans. Power Appar. Syst. **100**(11), 4665–4672 (1981)
2. Rasolonjanahary, J.L., Krahenbiihl, L., Nicolas, A.: Computation of electric fields and potential on polluted insulators using a boundary element method. IEEE Trans. Magn. **28** (2), 1473–1476 (1992)
3. Takuma, T.: Field behavior at a triple junction in composite dielectric arrangements. IEEE Trans. Electr. Insul. **26**(3), 500–509 (1991)

Chapter 2
Electric Field Behavior for a Finite Contact Angle

Introduction

When the cross-sectionally straight surface of a solid dielectric, or more generally the interface of two dielectrics, meets the surface of a plane conductor, the field strength is proportional to l^m near the point of contact, where l is the distance from the contact point. This field behavior means that the field strength theoretically becomes either infinitely high (singular behavior) or zero at the contact point, depending on the value of m.

As far as we know, J. Takagi was the first person to analyze the problem and to find the infinitely high field near the point of contact [1]. He applied an extended conformal transformation method, and also tried to obtain experimental confirmation based on the birefringence of solid dielectrics. In recent years, the field behavior near the contact point has become widely known as an important design parameter in the areas of waveguides and semiconductors. J. Meixner examined the phenomenon analytically by expanding the field as a power series in l [2].

With the intention of applying the field behavior to gas discharge, P. Weiss studied it numerically using the charge simulation method (CSM) in his dissertation, and named this behavior *Einbettungseffekt* (the German for embedding effect), after his experimental setup that consisted of a rod electrode embedded in an insulator column [3]. T. Takuma et al. analyzed the phenomenon by the CSM, the finite element method (FEM), and the analogue method using resistive paper, and named this occurrence of a zero or infinitely high electric field the Takagi effect, after the above-mentioned work of Prof. J. Takagi [4, 5]. K.J. Mürtz studied the field for the configuration shown in Fig. 2.1 in great detail by both analytical and numerical methods [6]. Furthermore, the effect of volume and surface conduction has been studied by T. Takuma, B. Techaumnat, and others under various conditions [7, 8].

T. Takuma and B. Techaumnat, *Electric Fields in Composite Dielectrics and their Applications*, Power Systems, DOI 10.1007/978-90-481-9392-9_2,

2.1 Analytical Treatment

2.1.1 Basic Field Behavior

First we will deal with an analytical study on the basic field behavior near a point of contact between a cross-sectionally straight dielectric interface (or solid dielectric surface) and the surface of a plane conductor. The configuration is represented in Fig. 2.1, where the straight interface of two dielectrics (with dielectric constants ε_A and ε_B) meets a plane conductor (electrode) at point P with a contact angle α in the ε_A side. In this chapter, α is given in radians unless otherwise stated.

When α is equal to $\pi/2$, the field is expected to exhibit no singularity near P. For $\alpha = \pi/2$ in Fig. 2.6 (shown later), for example, the field strength is E_0 or V/d (E_0 is a uniform field, V is the applied voltage, and d is the electrode separation) everywhere between the two electrodes. As explained in Section 1.4.2, on the other hand, the field strength in a very thin void (ε_A) lying inside another dielectric (ε_B) is nearly $\varepsilon_B/\varepsilon_A$ times that of the otherwise uniform field strength if the flat side is perpendicular to the direction of the external field. Thus, it is expected that the field strength approaches $\varepsilon_B/\varepsilon_A \times E_0$ (or $\varepsilon_B/\varepsilon_A \times V/d$) when α decreases to zero. In the past, it has falsely been supposed that when a void or a gap of ε_A is surrounded by ε_B, the field strength increases, at most, by a factor of $\varepsilon_B/\varepsilon_A$.

When we consider the field behavior in the close neighborhood of contact point P, the problem can be treated as two dimensional (2D). Analysis using the variable separation method is explained in detail in Section 8.1. Here we summarize the significant points of the results.

We can express the electric potential in the 2D polar coordinates (r, θ), as shown in Fig. 2.2, where the origin is the contact point P on the equipotential (or grounded) conductor surface. The potentials ϕ_A and ϕ_B in two dielectric media ε_A and ε_B are

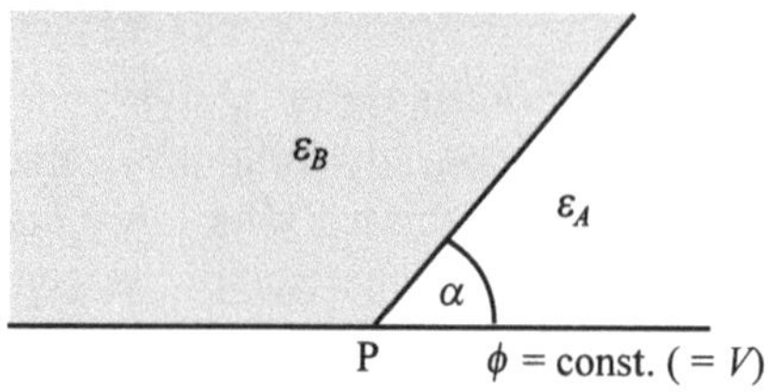

Fig. 2.1 Contact of a straight dielectric interface with a plane conductor (equipotential surface)

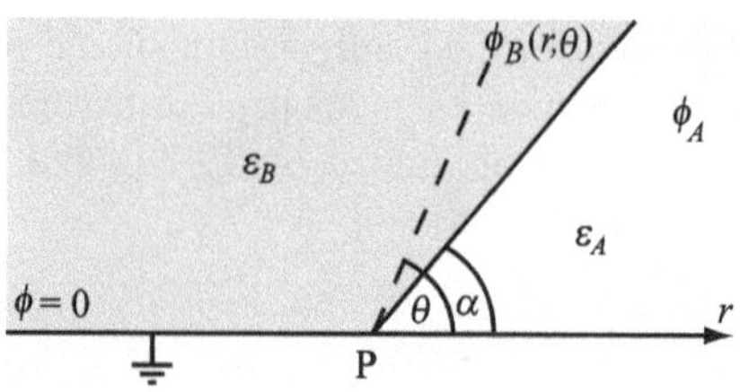

Fig. 2.2 Expression of the potentials in Fig. 2.1 in 2D polar coordinates

expressed as an infinite series of r^n multiplied by trigonometric functions, which satisfies Laplace's equation:

$$\phi_A \cong ar^n \sin(n\theta) \quad \text{and} \quad \phi_B \cong br^n \sin[n(\pi - \theta)], \tag{2.1}$$

where a and b are the constants independent of r and θ. It should be noted that the exponent n must be nonnegative to ensure a finite value of the potentials near the origin P, and that the smallest positive value of n is predominant near the point. From Eq. 2.1, field strength E is generally expressed near P as

$$E = K\,l^m, \tag{2.2}$$

where $m = n - 1$, l is the distance from the contact point, and K is a constant depending on the overall physical configuration and the ratio of the two dielectric constants. This expression means that at the point of contact, where $l = 0$, the field strength theoretically becomes either infinitely high or zero, unless $n = 1$.

Concrete values for n can be obtained from the following transcendental equation:

$$\varepsilon_B \tan(n\alpha) + \varepsilon_A \tan[n(\pi - \alpha)] = 0. \tag{2.3}$$

It can be easily understood that exchanging ε_A and ε_B in the equation has the same effect as changing the contact angle from α to $(\pi - \alpha)$, which is self-explanatory also from the geometry of the arrangement of Fig. 2.1.

Equation 2.3 has an infinite number of solutions for n. As explained above, however, the smallest positive value of n is predominant near the point of contact. Figure 2.3 represents such values of m $(= n - 1)$ as a function of contact angle α (in degrees) for ε_s $(=\varepsilon_B/\varepsilon_A)$ ranging from 2 to 10.

2.1.2 Minimum and Maximum Values of m in 2D Cases

It can be seen from Fig. 2.3 that m has a minimum between $\alpha = 0$ and $\pi/2$ and a maximum between $\alpha = \pi/2$ and π. The minimum and maximum values can be

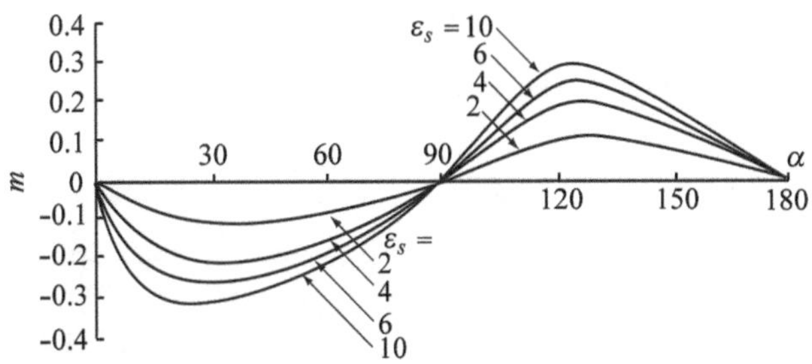

Fig. 2.3 Values of m as a function of α for several values of ratio ε_s $(= \varepsilon_B/\varepsilon_A)$ [4, 5]. © 1978 IEEE

obtained by differentiating Eq. 2.3 with respect to α and equating the result to zero. Combining the resulting equation with Eq. 2.3 gives

$$m = \pm\frac{1}{\pi}\ \arcsin\ \frac{\varepsilon_B - \varepsilon_A}{\varepsilon_B + \varepsilon_A}. \tag{2.4}$$

The contact angle which corresponds to the minimum (more important than the maximum) of m is

$$\alpha = \arcsin\sqrt{\frac{\varepsilon_A}{\varepsilon_A + \varepsilon_B}}\Big/\left(1 - \frac{1}{\pi}\arcsin\frac{\varepsilon_B - \varepsilon_A}{\varepsilon_B + \varepsilon_A}\right). \tag{2.5}$$

Equation 2.4 gives $|m| = 0.108$, 0.205, and 0.305 as maximum absolute values for ε_s ($= \varepsilon_B/\varepsilon_A$) $= 2$, 4, and 10, respectively. The corresponding angles are 0.690 (39.5°), 0.582 (33.4°), and 0.441 radian (25.2°), respectively.

The value of negative m decreases somewhat with increasing ε_s when α is kept constant, but the minimum value is only about –0.3 even for $\varepsilon_s = 10$. This means that the electric field strength increases relatively slowly as the point of contact is approached. When $\varepsilon_B \gg \varepsilon_A$, the minimum value of m falls to –1/2, while the maximum is 1/2 also for $\varepsilon_B \gg \varepsilon_A$. Thus the increase in field strength when approaching the contact point is slower than $l^{-1/2}$ for any 2D contact condition. As can be seen in Fig. 2.3, $m = 0$ both at $\alpha = 0$ and at $\alpha = \pi/2$, suggesting that the field does not exhibit the singularity of Eq. 2.2 in these cases.

2.1.3 Wedge-like Dielectric Interface Without a Contacting Plane Conductor

We explained in Section 1.4.1 that when a configuration has a grounded conductor plane, the identical field distribution can be realized by adding the mirror image (plane-symmetrical) configuration below the plane. This makes a wedge-like dielectric interface (protrusion or indentation) as shown in Fig. 2.4, which, without a contacting conductor, no longer forms a triple junction. The variable-separation method explained in Section 8.1.3 gives the following equation for determining exponent n in this case:

$$\left(\varepsilon_A^2 + \varepsilon_B^2\right)\sin A\sin(A - B) - 2\varepsilon_A\varepsilon_B\{1 - \cos A\cos(A - B)\} = 0, \tag{2.6}$$

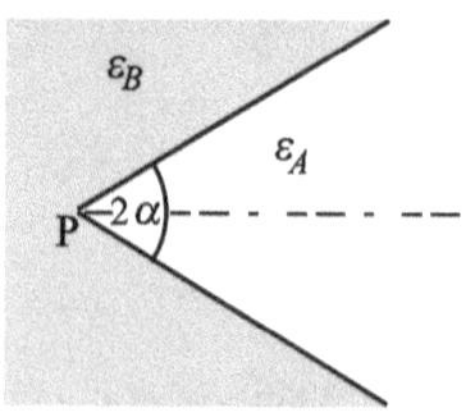

Fig. 2.4 Wedge-like straight dielectric interface of ε_A and ε_B

where $A = 2n\alpha$ and $B = 2\pi n$ for the configuration of Fig. 2.4. Further transforming this leads to the following simple expression:

$$\frac{\tan D}{\tan C} = \frac{\varepsilon_B}{\varepsilon_A} \quad \text{or} \quad \frac{\varepsilon_A}{\varepsilon_B}, \tag{2.7}$$

where $C = A/2 = n\alpha$ and $D = (A - B)/2 = n(\alpha - \pi)$. This expression becomes practically identical to Eq. 2.3 above, as expected.

2.1.4 Axisymmetric (AS) Case

As shown in Fig. 2.5, a cone-shaped protrusion (or projection) or void (if medium A is a gas or a vacuum) existing under otherwise uniform field E_0 makes an axisymmetric arrangement when E_0 is in the direction of the z-axis. The field behavior in this case is explained in Section 8.1.4. Potentials in the two dielectrics are given by an infinite series of r^n multiplied by Legendre functions in cylindrical coordinates (r, θ), the origin of which corresponds to apex P.

Similarly, as in 2D cases, the smallest positive value of n is predominant in the close vicinity of P, and field strength E is also expressed near P as

$$E = K\, l^m, \tag{2.8}$$

where l is the distance from P. The equation for determining the exponent n in this case is

$$\begin{aligned} &\varepsilon_B P_n(\cos\alpha)[\cos\alpha\, P_n(-\cos\alpha) + P_{n-1}(-\cos\alpha)] \\ &- \varepsilon_A P_n(-\cos\alpha)[\cos\alpha\, P_n(\cos\alpha) - P_{n-1}(\cos\alpha)] = 0, \end{aligned} \tag{2.9}$$

where P_n denotes an n-th order Legendre function. Figure 8.6 in Section 8.1 represents $m\ (= n - 1)$ computed at $\alpha = 0°$, $15°$, $30°$, etc. for several values of $\varepsilon_B/\varepsilon_A$.

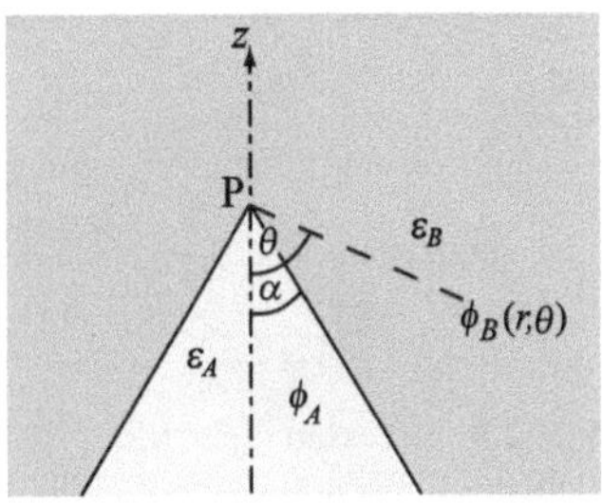

Fig. 2.5 Dielectric protrusion (or projection) or void existing under a uniform field in an axisymmetric (AS) configuration

2.2 Numerical Treatment

2.2.1 Dielectric Interface Between Parallel Plane Conductors

First we explain the field behavior in the configuration shown in Fig. 2.6 as a simple example with the contact angle being neither $\pi/2$ nor 0. In this figure, the surface of a solid dielectric (more generally, the interface of two dielectrics ε_A and ε_B) meets the surfaces of parallel plane conductors as a straight line on the sectional view. They thus make contact angles α and β ($=\pi - \alpha$) at the points of contact P and Q, respectively.

Contact conditions such as P or Q can occur in most solid dielectric supports because they always meet a conductor surface at a contact angle which is neither $\pi/2$ nor 0. Also, in the configuration of Fig. 2.7 (a void inside a solid dielectric with ε_A smaller than ε_B), the field behavior near P and P′ is similar to that near P in Fig. 2.6 because we can assume from its geometrical symmetry an equipotential surface passing through P and P′. The field in another practical configuration of Fig. 2.8 (a recessed dielectric slab) should also behave similarly near point P.

The field behavior of Eq. 2.2 holds only in the neighborhood of a contact point. We cannot determine the value of constant K by analytical methods, and must resort

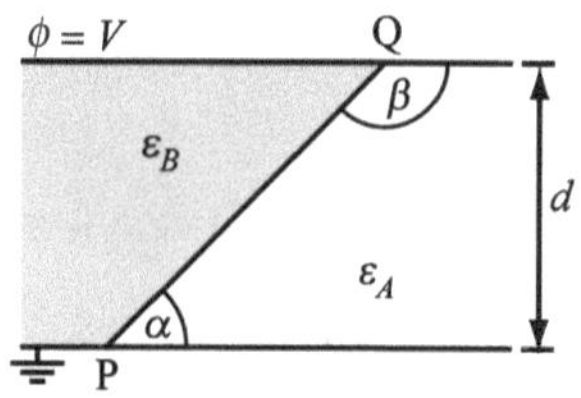

Fig. 2.6 Straight dielectric interface in contact with parallel plane conductors

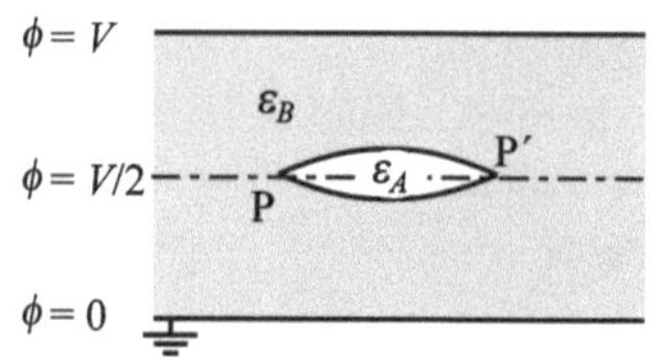

Fig. 2.7 Dielectric void inside another dielectric solid

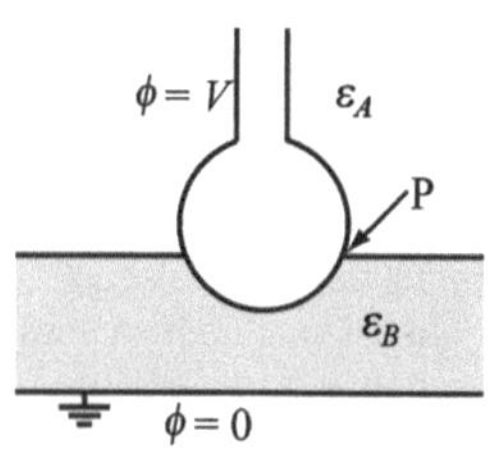

Fig. 2.8 Recessed dielectric slab

to a numerical method to analyze field distributions in more detail for such composite configurations encountered in practice, as shown in Figs. 2.6–2.8. Section 9.2 explains the calculation using the charge simulation method (CSM), and also presents a suitable arrangement of contour points (KPs, after the German word *Konturpunkt*) and fictitious charges (LADs, after the German word *Ladung*) for analyzing field behavior near the point of contact in composite dielectrics. In order to simulate the enhanced field behavior (the infinitely high field in particular) in a very narrow region near the point, KPs and LADs are placed there so that they become denser in geometric progression as the contact point is approached. This procedure is explained in Section 9.2.5; Fig. 9.6 represents one such arrangement used to analyze the field of Fig. 2.6.

Figure 2.9 shows the field strength on the ε_A and ε_B sides along the straight interface in Fig. 2.6, which was calculated by the CSM for $\varepsilon_s = 4$ and $\alpha = \pi/4$. In this figure, the ordinate is the absolute field strength E normalized by the average value (or uniform field without an interface) V/d, whereas the abscissa is the normalized distance l/d measured from the point of contact P or Q. The resulting characteristics appear linear when plotted on logarithmic scales in the range of l/d up to about 0.1, which confirms the relationship of Eq. 2.2 expressed as

$$E = K_1\,(V/d)\,(l/d)^m\,. \tag{2.10}$$

To put it concretely for the conditions given above, the field strength on the ε_A side along the interface of Fig. 2.6 for $\varepsilon_s = 4$ and $\alpha = \pi/4$ is

$$E = 1.35\,(V/d)(l/d)^{-0.194}\,. \tag{2.11}$$

For contact angle α smaller than $\pi/2$ in Fig. 2.6, the sign of exponent m is affected by the relative sizes of dielectric constants ε_A and ε_B in the following way:

$$m < 0 \text{ for } \varepsilon_B > \varepsilon_A, \text{ and } m > 0 \text{ for } \varepsilon_B < \varepsilon_A. \tag{2.12}$$

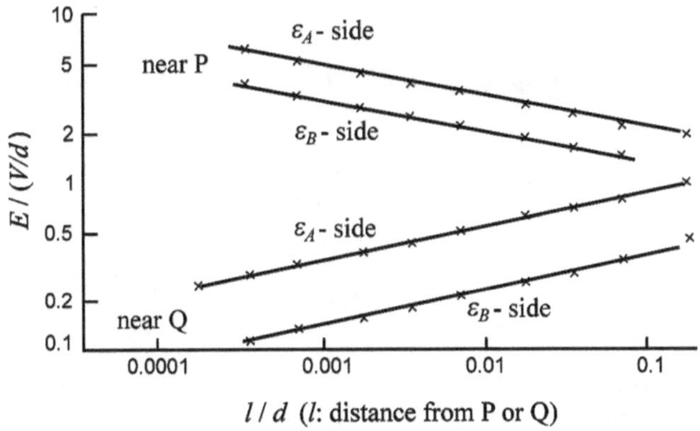

Fig. 2.9 Normalized field strength along the interface of Fig. 2.6 ($\varepsilon_s = 4, \alpha = \pi/4$) [4, 5]. © 1978 IEEE

The sign of m also changes according to whether α is larger or smaller than $\pi/2$. Thus when $\varepsilon_B > \varepsilon_A$ in Fig. 2.6, contact angles α and β produce an infinitely high electric field at point P and a zero field at point Q. This behavior is reversed when $\varepsilon_B < \varepsilon_A$, as can easily be seen from the inverse configuration of Fig. 2.6.

The field behavior of Eq. 2.2 has also been studied numerically by the finite element method (FEM) as well as by field mapping with resistive paper. The details are provided in an article by T. Takuma et al. [4]. Both results confirmed the linear relationship of field strength along the interface with the distance from a contact point on log–log scales, but these methods cannot explore minute field values in the way that the CSM can.

2.2.2 *Other Configurations*

A parametric computation on the effect of $\varepsilon_B/\varepsilon_A$ and α has been performed by the CSM for various configurations under a uniform field strength in two-dimensional (2D) and axisymmetric (AS) cases [4]. All the results confirm the linear characteristic of field strength relative to distance from a contact point on log–log scales near the contact point.

One example is the field distribution for an AS configuration of Fig. 2.10 under a uniform field. The calculation was done by the CSM for $\varepsilon_B/\varepsilon_A$ ranging from 1/6 to 6 with $\alpha = \pi/4$. Field strength E was expressed as Eq. 2.8 near apex P. Figure 2.11 shows the calculated results, where the field strength on the ε_A side is concretely expressed for $\varepsilon_B/\varepsilon_A = 1/6$ and $\alpha = \pi/4$ as

$$E = 1.36E_0(l/h)^{-0.34}. \tag{2.13}$$

Furthermore, it has been confirmed that the value of m from numerically computed field distributions (i.e., the slope of the lines in Fig. 2.11) agrees very well with the corresponding analytical value given by Eqs. 2.3 and 2.9 for each combination of $\varepsilon_B/\varepsilon_A$ and α in the 2D and AS cases, respectively.

2.2.3 *Effect of Right-Angled Contact (Curved Edge)*

The effect of curving an edge so that the dielectric interface makes contact with the conductor plane at a right angle has been studied for a 2D configuration modified

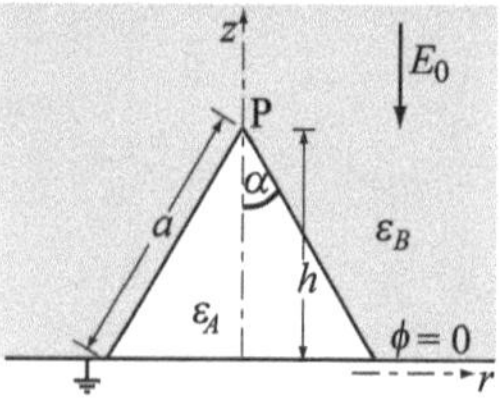

Fig. 2.10 Cone-shaped dielectric interface under a uniform field in an AS configuration

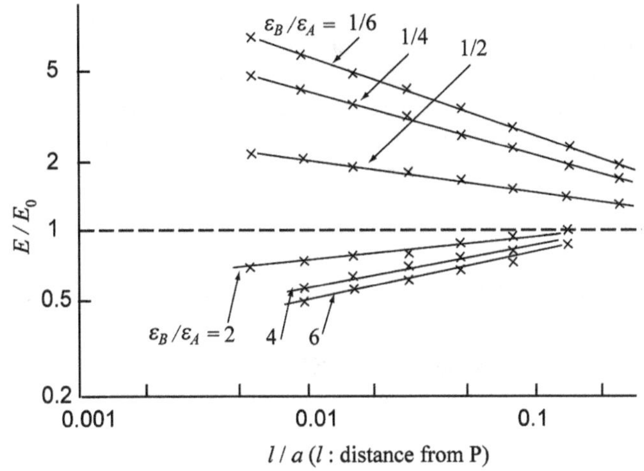

Fig. 2.11 Field strength near P on the ε_A side along the interface in Fig. 2.10 ($\alpha = \pi/4$) as calculated by the CSM [4]. © 1978 IEEE

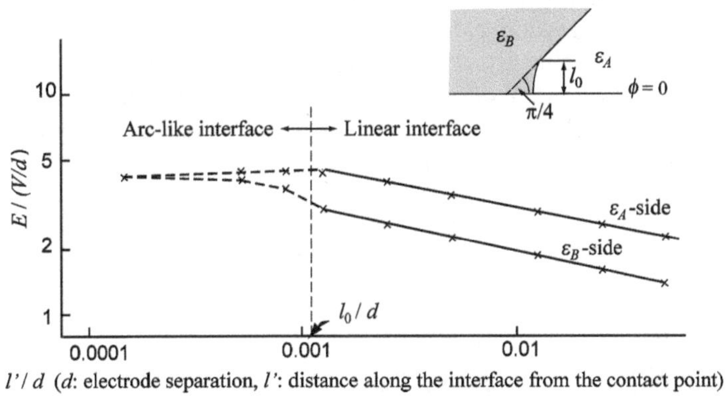

Fig. 2.12 Field strength along the interface for a configuration similar to Fig. 2.6 but with a curved interface near the contact point ($\varepsilon_s = 4$) [4]. © 1978 IEEE

from Fig. 2.6. As shown in the upper-right part of Fig. 2.12, the interface is straight everywhere except in the close vicinity of the conductor surface, where it is rounded so as to make an arc, thus meeting the surface at a right angle. The normalized distance l_0/d of the curved part is only about 0.001. The calculation was done by using the CSM.

Figure 2.12 shows the field strength on both the ε_A and ε_B sides along the interface. As the contact point is approached, the field strength on the ε_A side shifts from a linearly increasing characteristic to an almost constant value in the curved part on log–log scales. The field strength on the ε_B side, on the other hand, increases for a short interval with a higher slope than in the linear part, and then converges to the same value as on the ε_A side. It is concluded from this figure that the linearly increasing characteristic on log–log scales as shown in Fig. 2.9 is not due to the presence of a singularity at the contact point but rather to the overall straight contour of the interface. That is to say, the presence of a straight (but not right-angled) contour near the conducting plane gives rise to the field enhancement, depending on the distance of the straight part from the contact point.

2.3 Effect of Volume and Surface Conduction

2.3.1 *Complex Expressions for Fields*

The basic equations for so-called capacitive-resistive or mixed fields are explained in Section 1.3.1. We can calculate the fields by taking into consideration the true charge induced by surface or volume conduction. If volume conductivity is constant in a medium, no induced charge exists inside the medium, but only at its boundaries. Thus we can apply a boundary-dividing method to numerically analyze field behavior in this case. The important point in mixed fields is that the phasor notation (complex number expression) can be used to represent electric potential in a steady ac field of angular frequency ω (= $2\pi f$, where f is the corresponding frequency). As also shown in Eq. 1.15, the field including the effect of volume conduction can be simply expressed by substituting the following complex expression for the dielectric constant (relative permittivity) ε, as

$$\dot{\varepsilon} = \varepsilon + \frac{\sigma}{\mathrm{j}\omega\varepsilon_0} \tag{2.14}$$

where σ is the volume conductivity (= $1/\rho$, where ρ is the volume resistivity) and $\mathrm{j} = \sqrt{-1}$. The boundary conditions on the material interface, including the conductivity, are explained in Section 1.3.2.

2.3.2 *Basic Characteristics*

We consider the 2D arrangement of Fig. 2.2 in polar coordinates (r, θ), where medium B (dielectric constant ε_B) has volume conductivity σ and the interface has surface conductivity σ_s. In the close vicinity of contact point P, the potential takes a form similar to that in the previous sections,

$$\phi_A = Ar^z \sin z\theta \tag{2.15}$$

and

$$\phi_B = Br^z \sin z(\pi - \theta), \tag{2.16}$$

where $z = n + \mathrm{j}n'$ is a complex number having the smallest n that fulfills the following condition:

$$\varepsilon_A \cot z\alpha + \dot{\varepsilon}_B \cot z(\pi - \alpha) = \frac{\sigma_s}{\mathrm{j}\omega\varepsilon_0 r}(z - 1). \tag{2.17}$$

This equation can be derived by applying the boundary conditions to Eqs. 2.15 and 2.16 [8]. It is clear that by taking the gradient of ϕ_B or ϕ_A, the electric field becomes zero at the point of contact if $n > 1$, and infinitely high if $n < 1$, whereas n' contributes only to the phase shift of the potentials.

In the absence of surface conduction ($\sigma_s = 0$), the right-hand side of Eq. 2.17 vanishes and z can be solved numerically by using a suitable iterative method such as the modified Newton method [9]. On the other hand, if there is conduction along the interface, Eq. 2.17 implies that as P is approached,

$$\lim_{r \to 0} z = 1. \tag{2.18}$$

2.3.3 Effect of Volume Conduction

Figure 2.13 shows the 2D configuration corresponding to Fig. 2.6 where the interface of two materials A and B, assumed to be a gas or vacuum (ε_A) on the right hand side, and a solid (ε_B) on the left, meets parallel plane conductor surfaces with contact angles α and β at the points of contact P and Q, respectively ($\alpha + \beta = \pi$). Conduction in medium B is represented by volume conductivity σ and conduction on the interface by surface conductivity σ_s. Both σ and σ_s are assumed to be constant and uniform in the regions concerned. Here we consider a fixed power frequency (= 50 Hz) for the applied voltage, and investigate the behavior of field for various values of conductivity.

When medium B has uniform volume conductivity σ, the field strength increases near P with increasing σ, in contrast to the case with surface conduction, which will be explained later. With $\varepsilon_B = 4$ and $\varepsilon_A = 1$ for typical solid and gaseous insulating materials, Fig. 2.14 represents field strength E in the presence of volume conduction when σ_s is zero, where l is the distance along the interface from the contact point. The field, normalized by V/d, is taken on the A side at the material interface and is shown on log–log scales. This field behavior can be understood as a variant of the Takagi effect (embedding effect) in steady current fields in which volume

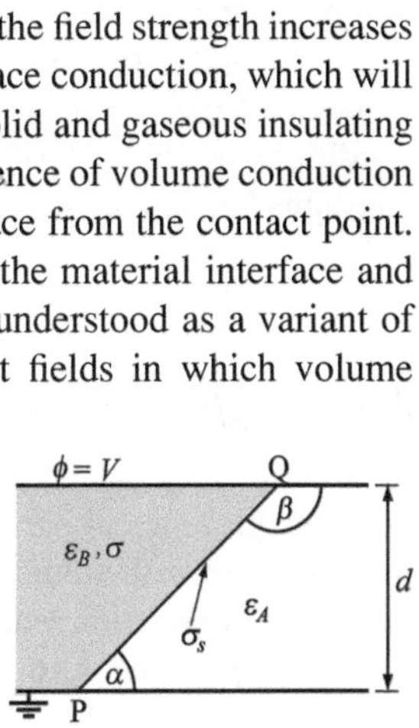

Fig. 2.13 Configuration as given in Fig. 2.6 with volume or surface conduction. Medium A is a gas (or a vacuum) and medium B is a solid

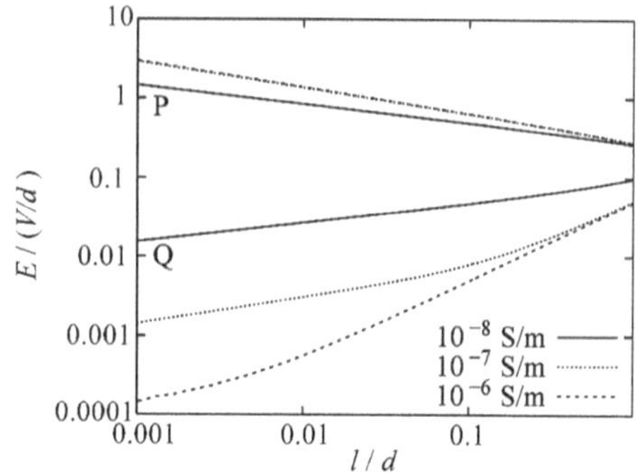

Fig. 2.14 Field distribution on the A (gas) side of material interface PQ of Fig. 2.13 with volume conduction for different values of volume conductivity σ ($\varepsilon_B/\varepsilon_A = 4$, $\alpha = \pi/4, f = 50$ Hz) [8]. © 2002 Elsevier Science B.V.

Table 2.1 Comparison of the analytical values of z determined from Eq. 2.17 and n (real part of z) obtained from the numerical results [8]. © 2002 Elsevier Science B.V.

σ (nS/m)	Analytical (z)	Numerical (n)
10	0.7613 + j0.0647	0.7613
100	0.6696 + j0.0200	0.6697
1,000	0.6667 + j0.0020	0.6667

conduction contributes to an increase in the complex dielectric constant in electric fields. E is either zero or infinitely high at the point of contact when the contact angle α on the A side is greater or smaller than $\pi/2$, respectively. When P or Q is approached, E increases or decreases more rapidly with distance from the contact point for higher conductivity values. From this rate of increase or decrease, we can calculate the real part n of z in Eqs. 2.15 and 2.16.

Values of n from numerical field calculations and from the solution of Eq. 2.17 are shown in Table 2.1; a good agreement was found between the analytical and numerical results. In the extreme case where σ is very high, z converges to real number n determined solely by the contact angle, i.e., n is the same as that for a very large real dielectric constant ε (for example, $n = 2/3$ for $\alpha = \pi/4$). In conclusion, σ has a similar effect to that of ε on the field behavior because the boundary conditions on the material interface take the same form, except that the dielectric constants become complex numbers.

2.3.4 Effect of Surface Conduction

With the presence of surface conduction, the electric field behavior differs greatly from the case with volume conduction. As explained in Section 2.3.1, the existence of surface conduction leads to Eq. 2.18 near point P, which creates a constant field from Eqs. 2.15 and 2.16. With z being unity, ϕ_A and ϕ_B near P become

$$\phi_A = ar\sin\theta \tag{2.19}$$

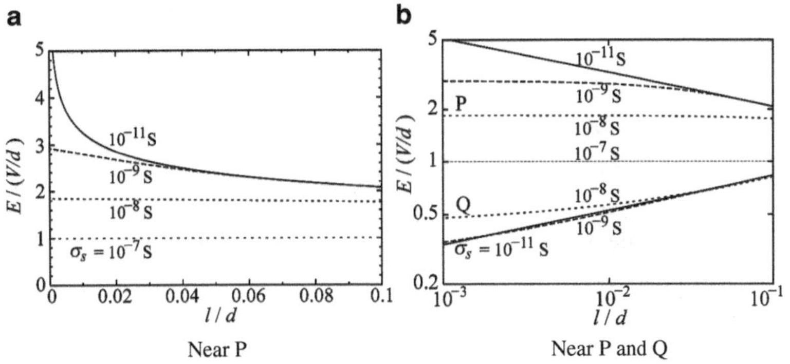

Fig. 2.15 Normalized electric field on the A (gas) side along the material interface of Fig. 2.13 with surface conduction ($\varepsilon_B/\varepsilon_A = 4$, $\alpha = \pi/4, f = 50$ Hz, $\sigma = 0$) [5, 8]. © 1991 IEEE, © 2002 Elsevier Science B.V.

and

$$\phi_B = br\sin(\pi - \theta). \tag{2.20}$$

These potentials fulfill the condition of potential continuity across the interface PQ. The electric field normal to the interface is then given by

$$E_{nA} = E_{nB} = a\cos\alpha, \tag{2.21}$$

which gives rise to a nonzero value on the left-hand side of the boundary condition (Eq. 1.21 in Chapter 1) equal to the divergence of the surface current density on the right-hand side. This is realized by the existence of surface conduction.

The effect of surface conduction is shown in the numerically calculated result of Figs. 2.15a and b which present the normalized field strength on the A side at the material interface in Fig. 2.13 (volume conductivity σ is zero). The calculation was performed by both the CSM using complex fictitious charges and the BEM. The frequency of the applied ac voltage was 50 Hz. Figure 2.15a uses linear scales and demonstrates that in the case of high conductivity or dc voltage, the field distribution is determined solely by the conductive component of impedance, thus resulting in a tangential component of $(V/d)\sin\alpha$ along the interface and a uniform total (absolute) field of V/d everywhere. On the other hand, Fig. 2.15b, which is presented on log–log scales, shows that the characteristic approaches a linear one with decreasing σ_s. The presence of surface conduction results in an almost constant field strength in the vicinity of contact points P and Q. The distance that the constant field extends from the contact point becomes larger with higher σ_s, as seen in Fig. 2.15b.

2.3.5 Approximate Evaluation of the Effect of Surface Conduction

The effect of surface conductivity σ_s can be evaluated by comparing the two parallel impedances per unit horizontal length in a corresponding interface region of Fig. 2.13 [10]. The conductive impedance Z_R and the capacitive impedance Z_C are, respectively,

$$Z_R = \frac{L}{\sigma_s},\; Z_C = \frac{1}{\omega C} = \frac{L \sin\alpha}{2\pi f \varepsilon_E \varepsilon_0 L \cos\alpha} \tag{2.22}$$

where L is the corresponding length along the interface, the total length of which is $d/\sin\alpha$. The equivalent dielectric constant ε_E was approximated by $\sqrt{\varepsilon_A \varepsilon_B}$. Thus,

$$\frac{Z_R}{Z_C} = \frac{2\pi f \varepsilon_E \varepsilon_0 L \cot\alpha}{\sigma_s}. \tag{2.23}$$

This equation can be used to roughly estimate the distance L_e of the interface (from the contact point) in which the surface conduction modifies the field to an almost constant value, as shown in Fig. 2.15b. The length L_e is directly proportional to σ_s and inversely proportional to frequency f and to ε_E.

The total parallel impedance Z is

$$Z = Z_C \Big/ \sqrt{1 + (Z_C/Z_R)^2}. \tag{2.24}$$

If we consider the case in which the contribution of Z_R amounts to about 10% of the total impedance, i.e., when Z/Z_C is 0.9, this condition leads to Z_R/Z_C being equal to about 2. Thus, equating Z_R/Z_C to 2, and with $\alpha = \pi/4, f = 50\text{Hz}, \varepsilon_A = 1$, and $\varepsilon_B = 4$, Eq. 2.23 becomes

$$L_e = 3.6 \times 10^8 \sigma_s \text{ (m)}. \tag{2.25}$$

The proportionality between L_e and σ_s is roughly substantiated in Fig. 2.15, but the values of L_e computed from Eq. 2.25 are a few times larger than the numerical results of Fig. 2.15, which were calculated for $d = 4$ m.

References

1. Takagi, J.: On the field at a tip of a conductor or dielectric. Waseda Denkikougakkai Zasshi (J. Electr. Eng. Dep., Waseda Univ.), 69–77, 103–110, and 139–147 (1939) (in Japanese)
2. Meixner, J.: The behavior of electromagnetic fields at edges. IEEE Trans Antennas Propag. **20** (4), 442–446 (1972), and Mittra, R., Lee, S.W.: Analytical techniques in the theory of guided waves, pp. 4–11. Macmillan, New York (1971)
3. Weiss, P.: Rotationssymmetrische Zweistoffdielektrika. Diss. Tech. Univ. Munich (1972) (in German), and Weiss, P.: Feldstärke-Effekte bei Zweistoffdielektrika, Proc. 1st ISH (Int. Symp. High Volt. Eng.), 73–79 (1972) (in German)
4. Takuma, T., Kouno, T., Matsuda, H.: Field behavior near singular points in composite dielectric arrangements. IEEE Trans. Electr. Insul. **13**(6), 426–435 (1978)

5. Takuma, T.: Field behavior at a triple junction in composite dielectric arrangements. IEEE Trans. Electr. Insul. **26**(3), 500–509 (1991)
6. Mürtz, K.J.: Das Hochspannungsfeld abgerundeter fiktiver Kanten. Diss. Kaiserlautern Univ. (1963) (in German), and Mürtz, K.J.: Das elektrische Feld abgerundeter fiktiver Kanten. etz Archiv **4**(1), 15–18 (1982) (in German)
7. Takuma, T., Kawamoto, T., Fujinami, H.: Effect of conduction on field behavior near singular points in composite medium arrangements. IEEE Trans. Electr. Insul. **17**(3), 269–275 (1982)
8. Techaumnat, B., Hamada, S., Takuma, T.: Effect of conductivity in triple-junction problems. J. Electrost. **56**, 67–76 (2002)
9. Kowalik, J., Osborne, M.R.: Method for unconstrained optimization problems. Elsevier, New York (1968)
10. Takuma, T., Kawamoto, T.: Effect of surface conductivity on the electric field at a triple junction in composite dielectric arrangements. The 1998 Annu. Meet. Rec. IEEJ. (Inst. Electr. Eng. Japan): No. 17 (1998) (in Japanese)

Chapter 3
Electric Field for a Zero Contact Angle (Smooth Contact)

Introduction

In the case of smooth contact with a zero contact angle between a solid dielectric and a surface, the field at the point of contact exhibits neither a singularity (an infinitely high value) nor a zero value, but it can be significantly high. For the simple case of a spherical or cylindrical conductor resting on an infinitely thick solid dielectric, the field strength at the contact point is expressed by a quadratic function of ε_s on the gas side and by a linear function of ε_s on the solid side. The ratio ε_s is equal to $\varepsilon_B/\varepsilon_A$, where ε_B and ε_A are, respectively, the dielectric constant, or relative permittivity, of a solid and a surrounding medium. This field behavior was derived analytically by the classical image method by T. Takuma and T. Kawamoto [1, 2]. However, there had been several papers dealing with the field behavior for a zero contact angle published earlier, and these are briefly mentioned here. Further details are provided by T. Takuma and T. Kawamoto [1].

J.H. Wensley and F.W. Parker calculated the field for a stress-relieving ring on a high-voltage cable by the finite difference method (FDM) early in 1956, but reported only its potential distribution on the paper surface rather than the field strength [3]. The simple configuration of a spherical conductor placed on a dielectric plate, as later shown in Fig. 3.8, was dealt with in three papers from the 1960s to the 1980s. K.C. Kao and T. Harker performed an analytical calculation which used an infinite series of Bessel functions to obtain the solutions to Laplace's equation in cylindrical coordinates [4]. In contrast, D.F. Binns and T.J. Randall applied the numerical method of the FDM [5]. The paper by Y.L. Chow et al. studied the behavior of conducting particles in solid dielectric-coated electrode systems, including a charged sphere under a uniform external field [6]. P. Weiss applied the charge simulation method (CSM) to the computation of electric fields in composite dielectric configurations in his dissertation, and one of his calculation examples was for a spherical electrode lying on a hemispherical solid dielectric [7]. K. Itaka's group analyzed the field behavior near a point of contact of solid insulating supports (spacers) in SF_6 gas-insulated cables, which will be explained in Section 5.2.4 [8, 9 in Chapter 5].

T. Takuma and B. Techaumnat, *Electric Fields in Composite Dielectrics and their Applications*, Power Systems, DOI 10.1007/978-90-481-9392-9_3,

With the development of high-speed, large-capacity computers and improved computational techniques, a number of papers have reported the field behavior near contact points in composite dielectrics, in particular, for practical equipment. However, most of these papers apply only to special configurations or conditions, and cover only a limited range of parameters. T. Takuma's group reported more general features of the field near a contact point analyzed by both analytical and numerical methods, and proposed a simple approximation relating the field strength at the point of contact to ε_s [1, 2]. B. Techaumnat et al. analyzed the effect of volume and surface conduction for zero-angle contact [8, 9]. They also reported the analysis of the electric field and force on a conducting sphere lying on a solid dielectric under a uniform field [10].

In this chapter, we explain the field behavior for configurations that are as simple as possible so as to clarify the fundamental characteristics. The results in more practical cases are dealt with later in Chapters 5–7. Angles are given in radians in this chapter.

3.1 Stressed Conductor in Contact with a Solid Dielectric

3.1.1 Field Strength at a Point of Contact

Smooth contact with a zero contact angle means that a conductor and a solid dielectric have a common tangent at the point of contact, forming no sharp edge. Under this condition, the conductor or the solid (or both) must be curved.

The simplest configuration with a round conductor is shown in Fig. 3.1, where a conductor rests on the plane surface of an infinitely thick solid dielectric of dielectric constant ε_B. There exist basically two different cases for this configuration:

(a) The conductor is stressed or charged to a fixed potential [1, 2], and
(b) The conductor is at a floating potential under an externally applied field [10].

Case (a) is explained in this section, while case (b) is treated as an uncharged conductor under a uniform field in Section 3.2. For the general case of a charged,

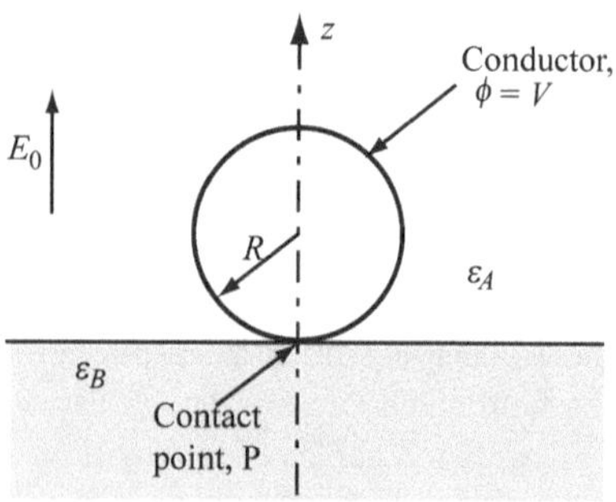

Fig. 3.1 Configuration of a spherical (or cylindrical) conductor resting on a dielectric plane

electrically floating conductor under an external field, the field is simply obtained by superposing the fields from cases (a) and (b).

The dielectric constant of the upper half space (usually a gas or a vacuum) is ε_A. The conductor is a sphere (point contact) for the axisymmetric (AS) condition, while it is a cylinder (line contact) in the two-dimensional (2D) condition. For AS cases, the field behavior in Fig. 3.1 can be analytically computed by the image charge method, in which an infinite number of image charges are located repetitively inside the sphere and the solid dielectric.

The concrete computation procedure is explained in Section 8.2.1. Here we summarize the significant points of the results. For this configuration, radius R is merely a proportionality constant or a normalizing factor, and the only effective parameter is the solid-to-gaseous ratio of the dielectric constants $\varepsilon_s = \varepsilon_B/\varepsilon_A$. When a spherical conductor possesses potential V in the absence of an external field in the AS configuration, the field strength at the point of contact P is expressed as

$$E_{CA} = \frac{(\varepsilon_s + 1)\varepsilon_s}{2} E_1 \text{ on the } \varepsilon_A \text{ side, and} \tag{3.1}$$

$$E_{CB} = \frac{E_{CA}}{\varepsilon_s} = \frac{\varepsilon_s + 1}{2} E_1 \text{ on the } \varepsilon_B \text{ side.} \tag{3.2}$$

E_1 is the field strength at the corresponding point (i.e., the lowest point on the sphere surface, although, in fact, the field strength is the same everywhere on the surface) for $\varepsilon_A = \varepsilon_B$, i.e., when the conductor exists isolated inside an infinite space of ε_A. E_1 is equal to V/R in the AS case. It is to be noted that E_{CA} is only a theoretical or a limiting value extrapolated to the contact point, where the ε_A gap disappears. On the other hand, E_{CB} does actually exist.

A similar repetitive image charge method can be applied to analyze the behavior in the 2D case where a cylindrical conductor rests on the solid dielectric plane. However, the field strength on the cylinder surface at a fixed potential theoretically becomes zero for an infinitely thick solid dielectric. Therefore, we analyze the field strength for finite thickness D as shown later in Fig. 3.8. For sufficiently large thickness D, E_1 for $\varepsilon_A = \varepsilon_B$ is approximated with the value given below in Eq. 3.3 in the 2D case. As explained in the computation process of Section 8.2.2, Eqs. 3.1 and 3.2 also apply to the 2D case.

3.1.2 Field Behavior near the Point of Contact

In order to clarify the degree of field enhancement near a contact point, we explain the field behavior in relation to the distance from point P. Unlike the field strength at a contact point composed of only one component vertical to the plane, the field has two components near P in both the AS (r- and z-components) and 2D (x- and y-components) cases. The analytical expressions of these components are given in a report by T. Takuma and T. Kawamoto [11].

Figures 3.2a and b show the tangential (or horizontal) field components E_t on the ε_A side along the plane surface (dielectric interface) versus the normalized distance from P in the AS (r-component) and 2D (x-component) configurations of Fig. 3.1, respectively. The vertical axis is normalized by the field strength on the conductor when $\varepsilon_s = 1$, as given in Eq. 3.3. As mentioned above, D in the equation means a sufficiently large thickness of the solid.

$$E_1 = V/R \text{ in AS cases, and}$$

$$E_1 = V/\ln(2D/R + 1) \text{ in 2D cases.} \tag{3.3}$$

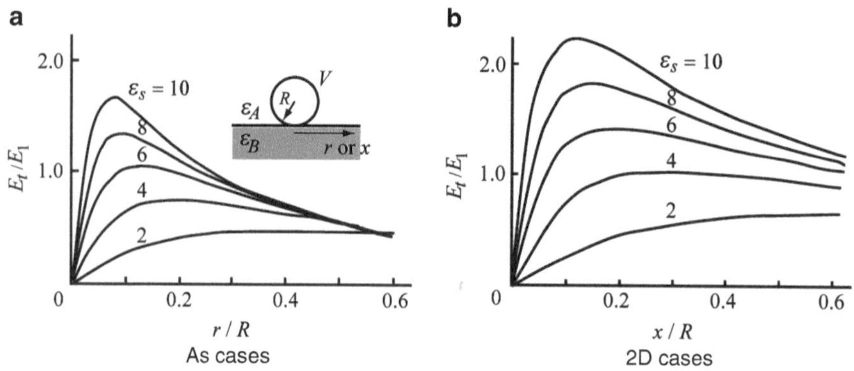

Fig. 3.2 Tangential field component on the ε_A side along the dielectric interface relative to the normalized distance from contact point P [11]

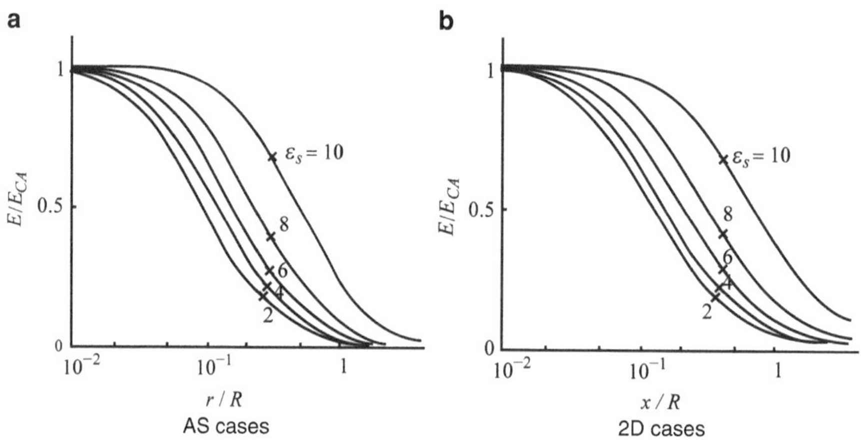

Fig. 3.3 Total (absolute) field strength on the ε_A side along the dielectric interface relative to the normalized distance from P. *Crosses* indicate the point where the field is ε_s times the value E_1 of Eq. 3.3 [11]

Moving away from P, the tangential component increases from zero at P, and then after a maximal value gradually decreases toward zero again. The position of the maximum field becomes closer to the contact point P with increasing ε_s.

In contrast, the vertical component is similar to the total (absolute) field strength represented in Figs. 3.3a and b. The field in these figures is normalized by E_{CA}, the value at contact point P. The graphs plotted in semi-logarithmic scales indicate the relatively rapid decrease of the high contact field with increasing distance from P. Cross marks (×) on the curves show the point where the field is ε_s times the value E_1 of Eq. 3.3. This point lies in a comparatively narrow range of distances, i.e., between $0.2R$ and $0.5R$ from P, and hardly changes with ε_s.

3.1.3 Conductor Separated from a Dielectric Plane

The field enhancement near the contact point markedly diminishes when a conducting sphere or cylinder is separated from the dielectric plane, as shown in Fig. 3.4, where the separation distance is S. The image charge method with an infinite number of point images can also be applied to this configuration, more details of which are given in Section 8.2.3.

The maximum field appears at point M (the lowest point) of the conductor facing the plane. Figures 3.5a and b represent the field strength E_M on the ε_A side at M in relation to S in AS (sphere-to-dielectric) and 2D (cylinder-to-dielectric) conditions, respectively. E_M is normalized by E_{CA}, the contact point strength of Eq. 3.1, and S by R. It is interesting to note that the dependence of E_M on S/R is similar to that of the field on r/R or x/R shown in Fig. 3.3. Cross marks (×) on the curves show the point where the field is ε_s times the value E_1 of Eq. 3.3. As was the case for the cross marks in Fig. 3.3, the points also exist in a relatively narrow range of S, e.g., within $0.045R$ and $0.07R$ in the AS cases, and are almost independent of ε_s.

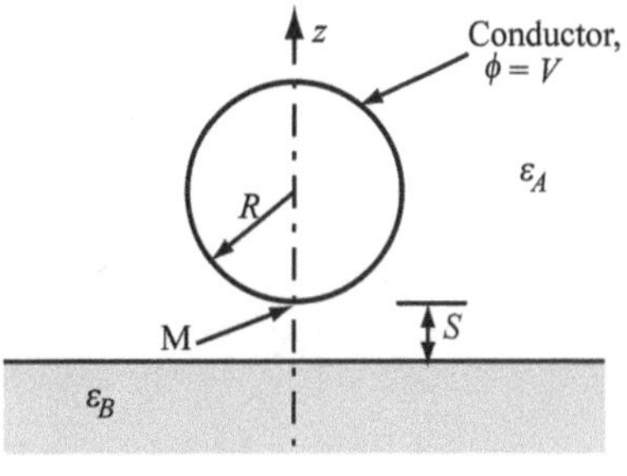

Fig. 3.4 Conductor existing above a dielectric plane with separation S

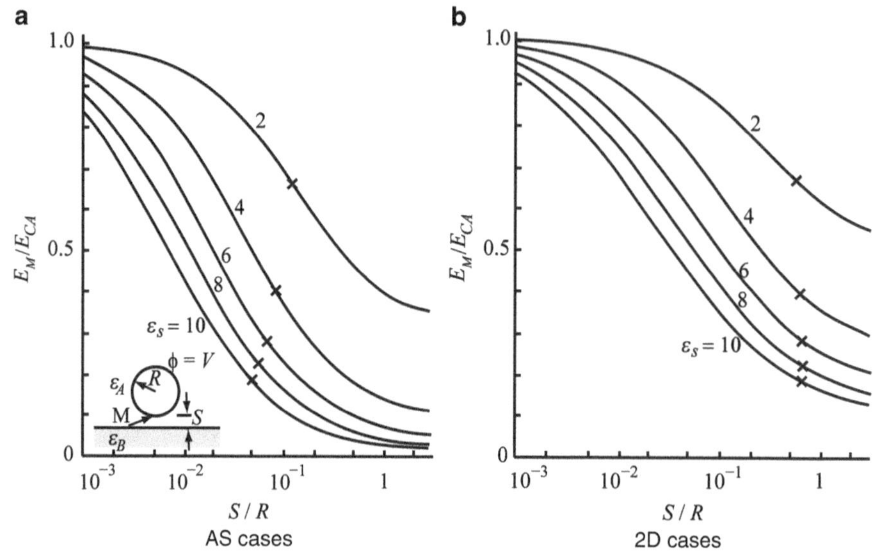

Fig. 3.5 Maximum field strength on the ε_A side for the configuration shown in Fig. 3.4. *Crosses* indicate the point where the field is ε_s times the value E_1 of Eq. 3.3 [11]

3.2 Uncharged Spherical Conductor Under a Uniform Field

3.2.1 Expression for Contact-Point Field

If the sphere in the configuration shown in Fig. 3.1 is an electrically floating conductor and is subjected to uniform field E_0, a more complicated image charge method is applied [10]. The detailed calculation technique is described in Section 8.3. From that analysis, the contact-point electric field is expressed as

$$\frac{E_{CA}}{E_0} = 1 + \frac{\varepsilon_s}{2}\left\{3 + \varepsilon_s - \alpha_P \frac{(\varepsilon_s + 1)^2}{(\varepsilon_s - 1)}\left[\ln \frac{(\varepsilon_s + 1)}{2} - \beta_P\right]\right\} \tag{3.4}$$

where α_P and β_P are functions of $P = (\varepsilon_s - 1)/(\varepsilon_s + 1)$. For the range $1 \leq \varepsilon_s < \infty$, $0 \leq P < 1$. α_P and β_P are defined as

$$\alpha_P = \sum_{k=1}^{\infty} S_k; \quad S_k = -\sum_{l=0}^{k-1} \frac{P^{k-l}}{k - l + 1} S_l \tag{3.5}$$

where $S_0 = 1$, and

$$\beta_P = \sum_{j=1}^{\infty} \frac{P^j}{j^2}. \tag{3.6}$$

Although we cannot write these parameters in a closed form, the summations converge well if P is not close to unity, which is true for most dielectrics in practical use. Numerically computed values of α_P and β_P are given in Fig. 3.6. It can be seen from the figure that $\alpha_P = 1$ at $P = 0$ and decreases with increasing P. The decrease of α_P is roughly linear for small P, but becomes steeper at larger P. In contrast, β_P increases with P and converges to the Riemann zeta function $\zeta(2) = \pi^2/6$ when P approaches unity.

3.2.2 Comparison of Contact-Point Fields

The solid and dashed lines in Fig. 3.7 present the contact-point electric field E_{CA} (normalized by E_0) on the ε_A side at a floating potential under E_0 and that on a sphere at fixed potential V (without external field), respectively. For comparison, the potential V in the latter case is chosen so that the magnitude of E_{CA} for $\varepsilon_B = \varepsilon_A$ is the same as that on the sphere under E_0. That is to say, V is equal to $3RE_0$. It should be noted that because the gap length is zero at the contact point, E_{CA} represents the upper bound (limiting value) of electric field in a small gap near the contact point. On the ε_B side, the field, equal to E_{CA}/ε_s, actually exists, as already mentioned.

It can be clearly seen from Fig. 3.7 that the ratio E_{CA}/E_0 increases as ε_s increases. The figure also suggests that the field will be infinitely high if $\varepsilon_s = \infty$. In usual

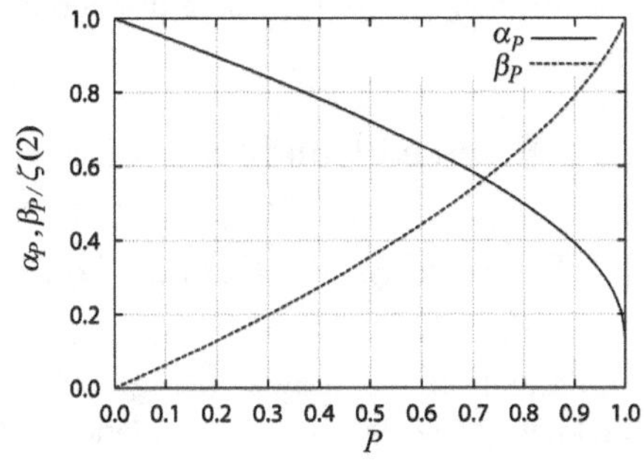

Fig. 3.6 Values of α_P and β_P expressed by Eqs. 3.5 and 3.6 for the contact-point electric field [10]. © 2005 Elsevier Science B.V.

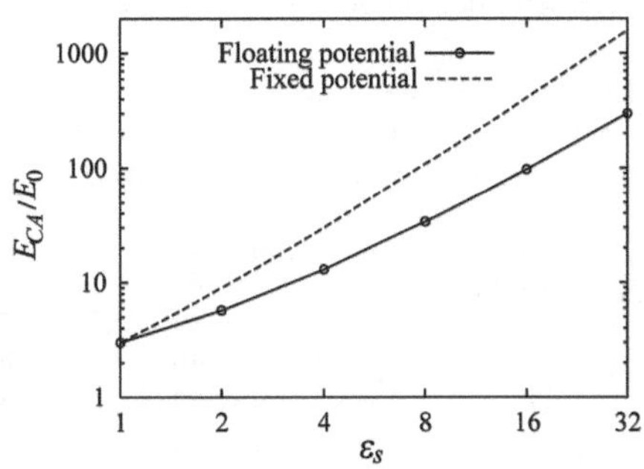

Fig. 3.7 Comparison between the contact-point electric field on a sphere at fixed potential V and that on an electrically floating sphere under uniform field E_0 [10]. © 2005 Elsevier Science B.V.

electric field distributions, the field for $\varepsilon_s = \infty$ corresponds to that for a conducting solid, or a conductor. However, in contact field problems, $\varepsilon_s = \infty$ is not equivalent to the case for a conductor. The solid still has no conductivity, so that no charge transfer occurs between the sphere and the solid. This corresponds to the situation in which an infinitesimally small gap exists between them. We can see from Fig. 3.7 that, for $\varepsilon_s > 1$, E_{CA} on a conducting sphere at a fixed potential is higher than that on a sphere under E_0. The difference in the corresponding E_{CA} values becomes greater as ε_s increases.

3.2.3 Approximate Expression

For an electrically floating conducting sphere, the contact-point electric field E_{CA} can be exactly computed from Eq. 3.4 by using α_P and β_P given in Eqs. 3.5 and 3.6. However, the following simple formula has been derived for E_{CA}, based on curve fitting:

$$E_{CA} \approx 0.696\left(\varepsilon_s + \sqrt{\frac{7}{4}}\right)^{\sqrt{3}}. \tag{3.7}$$

This equation approximates E_{CA} well over the range of $\varepsilon_s = 1$–64. Table 3.1 compares the contact-point field obtained from the approximation of Eq. 3.7 and the analytically computed value, showing that the error from the approximation is smaller than 3% for the range of ε_s up to 64.

3.3 Stressed Conductor on a Solid Dielectric of Finite Thickness

3.3.1 Field Strength at a Contact Point

Configurations as shown in Fig. 3.8, in which the solid dielectric has a finite thickness of D on a grounded plane conductor, are often utilized, e.g., in experiments for evaluating the electrical strength of a solid insulating material. Compared to the case of a dielectric with infinite thickness described above in Section 3.1.1, D/R becomes an additional parameter governing the field behavior.

Table 3.1 Comparison of the exact contact-point electric field formulation with the approximation given in Eq. 3.7 [10]. © 2005 Elsevier Science B.V.

ε_s	E_{CA}/E_0		Difference (%)
	Exact	Eq. 3.7	
1	3.000	2.996	0.133
2	5.710	5.571	2.435
4	12.94	12.60	2.640
8	33.68	33.26	1.237
16	96.80	97.27	0.478
32	298.2	302.0	1.286
64	964.3	969.2	0.504

Figures 3.9a and b represent field strength E_{CA} at the point of contact P on the ε_A side as a function of D/R for AS and 2D cases, respectively. The ordinate E_{CA} is normalized by the field strength E_{C1} at the corresponding point for $\varepsilon_s = 1$, i.e., when a spherical or cylindrical conductor faces a grounded plane separated by D in a single dielectric medium. E_{C1} is different from E_1 given in Eq. 3.3 due to the presence of the grounded plane at a finite distance. These figures show that the field strength rapidly increases with D and then levels off to the limiting value given by Eq. 3.1. For AS configurations, in particular, E_{CA} is already close to the limit when $D/R = 1$.

Figure 3.10 represents the field strength on the ε_A side along the dielectric interface for $D = 2R$, analogously to Fig. 3.3 for a dielectric of infinite thickness, but the abscissa is given in a linear scale in this figure.

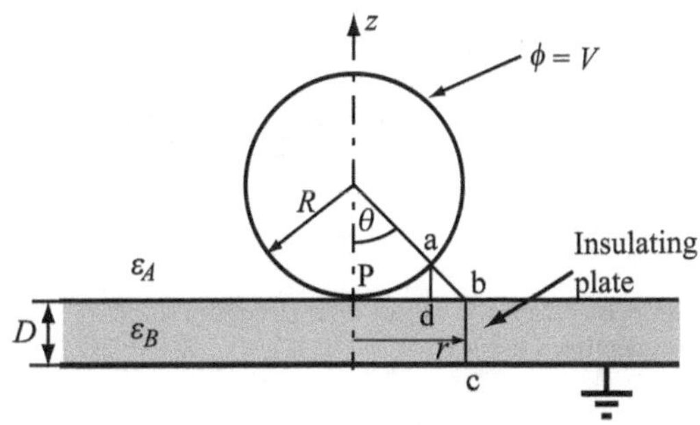

Fig. 3.8 Sphere-to-dielectric plane (AS) or cylinder-to-dielectric plane (2D) configuration with the solid dielectric having finite thickness

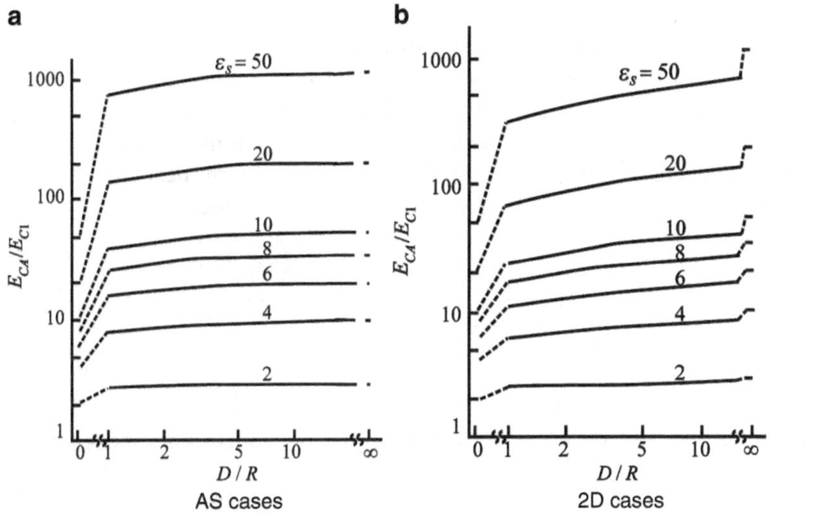

Fig. 3.9 Field strength at contact point P on the ε_A side for AS (sphere-to-dielectric plane) and 2D (cylinder-to-dielectric plane) configurations. Modified from Figs. 6 and 7 in [1, 2]. © 1984 IEEE

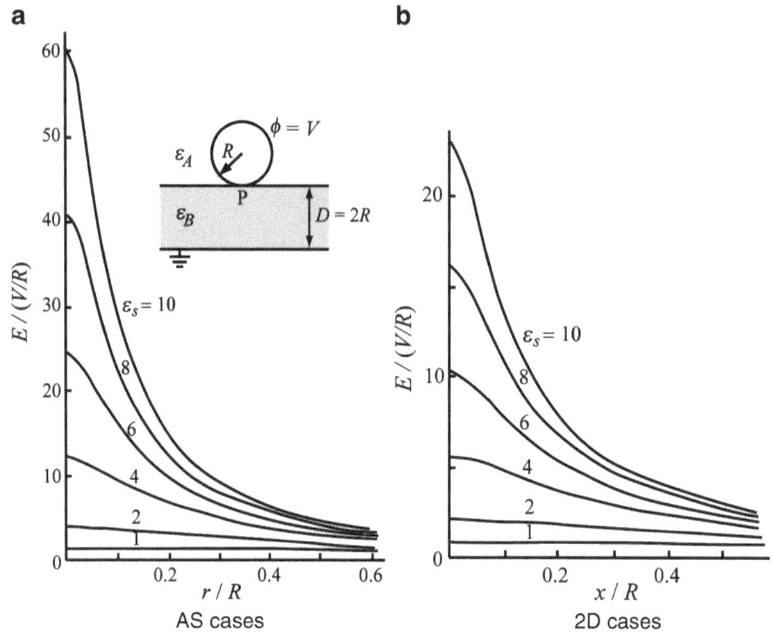

Fig. 3.10 Total (absolute) field strength on the ε_A side along the dielectric interface relative to distance from point P ($D = 2R$) [1]. © 1984 IEEE

3.3.2 *Approximate Treatment Based on Series Capacitance*

The following approximations are sometimes made, e.g., to estimate the partial discharge inception voltage near a contact point in the configuration given in Fig. 3.8. Namely, the field strength is evaluated by dividing the applied voltage between the gas and the solid dielectric as a capacitive impedance in series. The approximation that the field distribution is almost uniform near the contact point roughly gives the following field strength E_A on the ε_A side:

$$E_A = \frac{V}{(\overline{\mathrm{ab}} + \overline{\mathrm{bc}}/\varepsilon_s)} = \frac{V}{\sqrt{R^2 + r^2} - R + (D/\varepsilon_s)}. \tag{3.8}$$

Points a, b, and c (also d used in Eq. 3.9) are indicated in Fig. 3.8. In the close vicinity of the contact point, i.e., when the distance $\overline{\mathrm{ab}}$ is small, this expression is almost the same as

$$E_A = \frac{V}{(\overline{\mathrm{ad}} + D/\varepsilon_s)} = \frac{V}{R - \sqrt{R^2 - r^2} + (D/\varepsilon_s)}, \tag{3.9}$$

and can be furthermore approximated with the following simpler expression:

$$E_A = \frac{V}{(r^2/2R) + (D/\varepsilon_s)}, \text{ and } E_B = E_A/\varepsilon_s. \tag{3.10}$$

The values generated by these expressions greatly differ from the numerically computed results in several respects, among which the most significant are as follows:

1. The field strength at the contact point on the ε_A side, E_{CA}, is much higher in Fig. 3.9 than the corresponding value $\varepsilon_s V/D$ derived from the approximate equations (Eqs. 3.8–3.10) for $r = 0$.
2. E_B on the ε_B side decreases with increasing ε_s in the approximations, whereas in the numerical results, not only E_{CA} but also E_{CB} increases with ε_s, as suggested by Eq. 3.2.

3.3.3 Field Behavior for Small *D*/*R*

The field distribution near a contact point has been calculated for small values of D/R by the re-expansion method (Section 8.4) for the sphere-to-dielectric plane (AS) configuration of Fig. 3.8. The conditions used in the calculations were $D/R =$ 0.05, 0.1, 0.2, 0.5, 1, 2, and ε_s ($=\varepsilon_B/\varepsilon_A$) = 2, 4, 6, 10.

Figures 3.11a and b show the field strength (total or absolute value E_{tot}) along the dielectric interface for $D = 0.05R$ on the ε_A side and on the ε_B side, respectively. Both the vertical and horizontal axes have linear scales in these figures. The field is

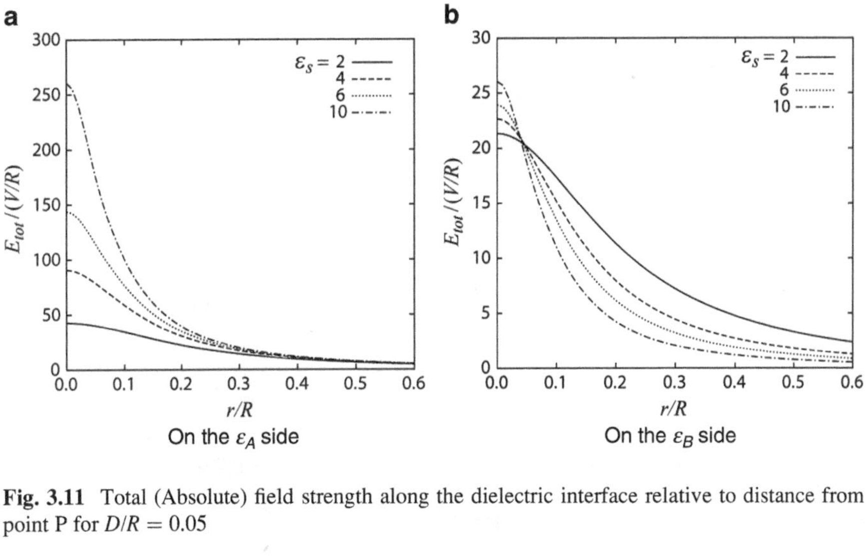

Fig. 3.11 Total (Absolute) field strength along the dielectric interface relative to distance from point P for $D/R = 0.05$

normalized by V/R, and the distance r from the contact point is normalized by R. The corresponding results are given in Fig. 3.12 for $D = 0.2R$ and in Fig. 3.13 for $D = R$. The field strength at the contact point on the ε_A side increases rapidly with ε_s and also with decreasing D/R. However, the latter characteristic, i.e., its dependence on D/R, is affected by the presence of the grounded plane: the shorter the distance to the grounded plane, the higher the field strength, even when $\varepsilon_s = 1$. If we normalize the field strength by V/D, it increases with increasing D/R, which means a higher field enhancement for larger D/R of the solid dielectric. Moving away from the contact point, the field strength for small D/R decreases more rapidly for higher

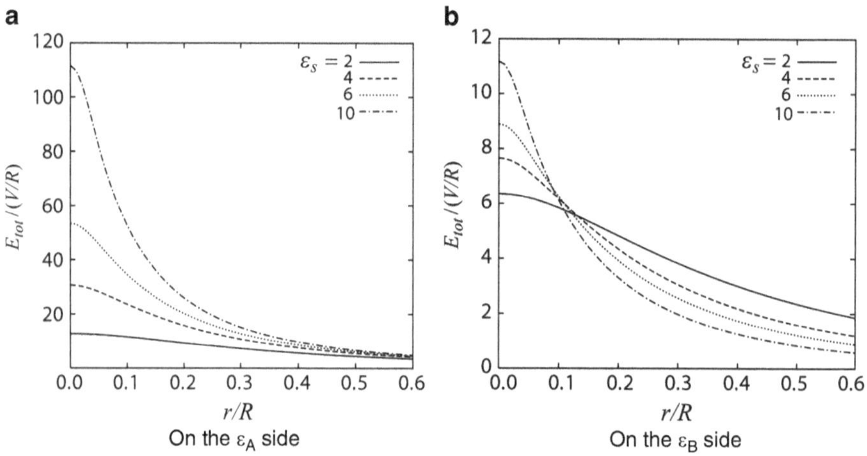

Fig. 3.12 Total (absolute) field strength for $D/R = 0.2$

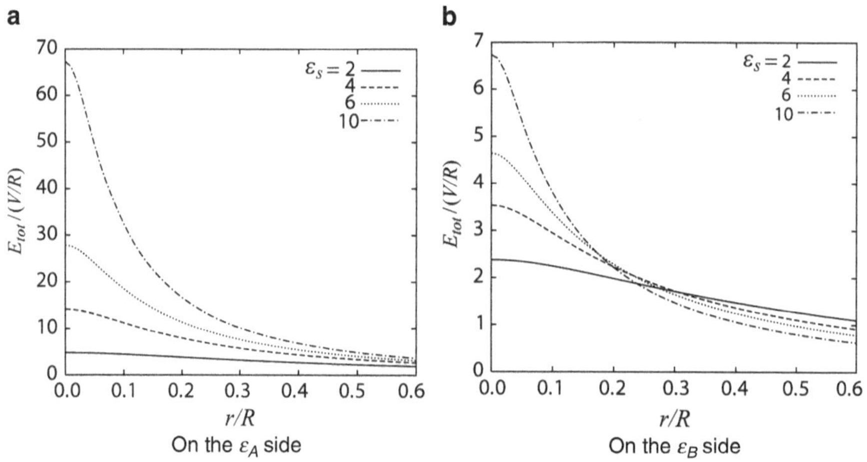

Fig. 3.13 Total (absolute) field strength for $D/R = 1$

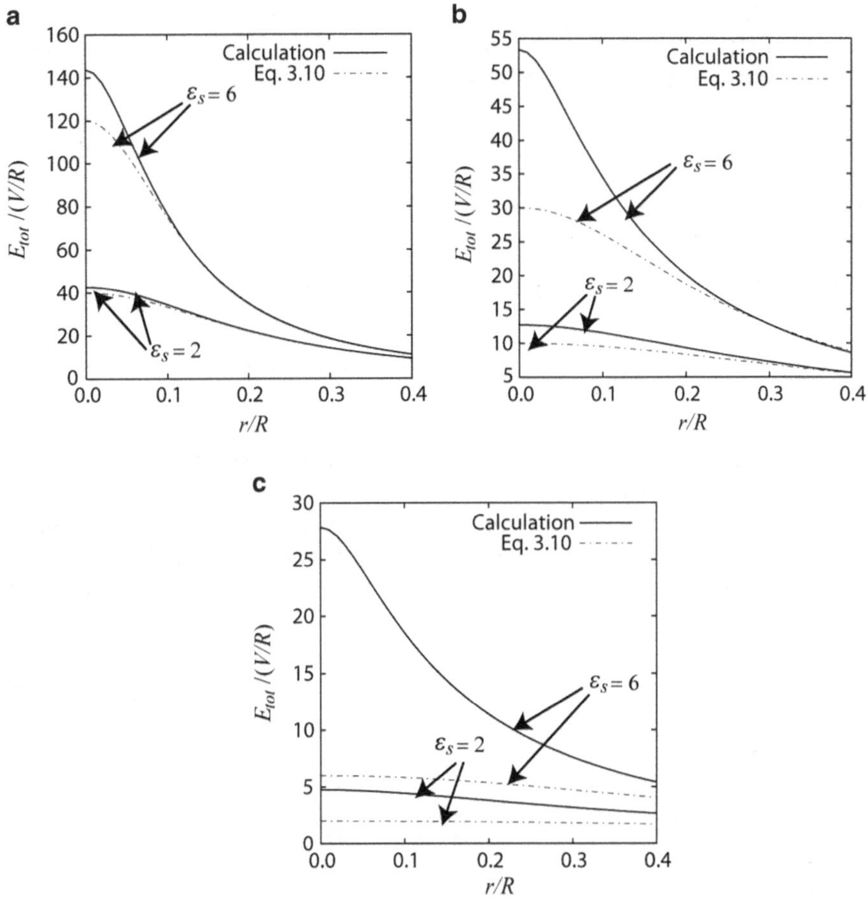

Fig. 3.14 Comparison of field strength on the ε_A side at the dielectric interface calculated by the re-expansion method with that given by Eq. 3.10

ε_s, analogously to Fig. 3.10a for $D = 2R$. The field strength on the ε_B side, on the other hand, is higher near a contact point for higher ε_s, but this tendency is reversed at a certain value of r, resulting in higher field strengths for lower ε_s.

The calculated field distribution on the ε_A side is compared with the approximation of Eq. 3.10 in Figs. 3.14a, b, and c for $D = 0.05R$, $0.2R$, and R, respectively. These results demonstrate that the field strength on the ε_A side differs from that generated using Eq. 3.10 at the point of contact, but with increasing r, it approaches and eventually coincides with the approximation. The coincidence is better when D/R is smaller. In conclusion, the field for small D/R behaves near a contact point similarly to that for larger or infinite D, as explained above, but moving away from the contact point, it approaches the characteristic of a series capacitance consisting of gas and solid dielectric gaps.

Figure 3.15 presents E_{CB} (contact-point field on the ε_B side) normalized by V/R versus ε_s for ε_s values up to 10 and for D/R ranging from 0.05 to 2. This characteristic is discussed later in Section 3.4.4.

3.4 Other Basic Configurations

3.4.1 Dielectric Cylinder Under a Uniform Field

An analytical solution can be obtained for the two-dimensional (2D) configuration of a dielectric cylinder lying on a conducting plane under a uniform electric field (Fig. 3.16). The computation is done by applying an infinite number of dipole images, which is not applicable to the axisymmetric (AS) situation of a dielectric sphere resting on a conducting plane. The electric field on the ε_B side of the cylinder surface is presented in Fig. 3.17 for ε_s up to 100. The maximum field strength is equal to the external field E_0 in any case, and this appears at the contact point. This means that the contact-point field E_{CA} on the ε_A side is intensified by the factor ε_s.

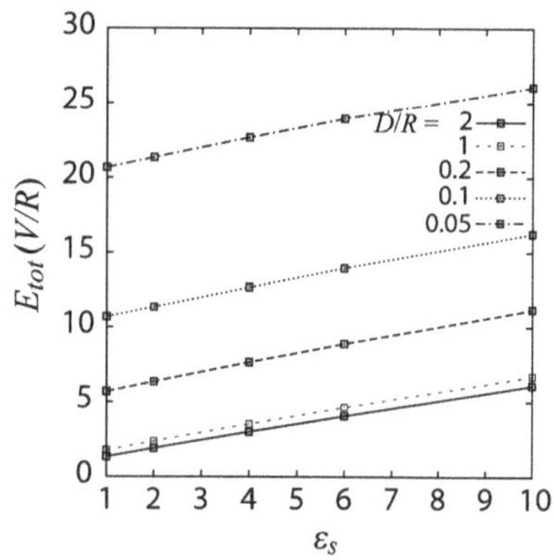

Fig. 3.15 Normalized field strength on the ε_B side at a point of contact for small D/R values for the configuration given in Fig. 3.8

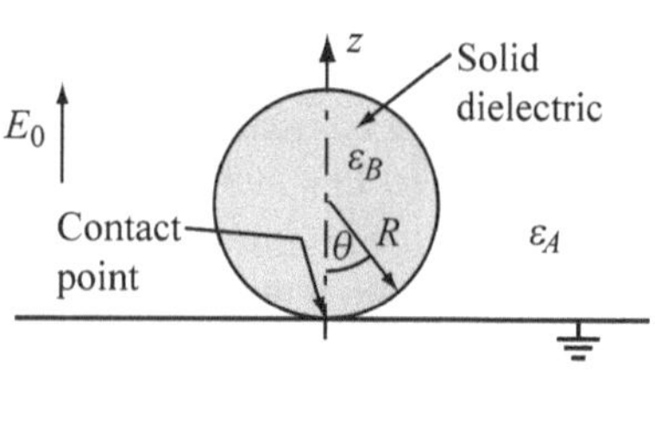

Fig. 3.16 A dielectric cylinder lying on a grounded conductor plane under an external uniform electric field

3.4.2 *Other Simple Configurations*

We here deal with a few simple configurations having a zero contact angle for which numerical approaches are more suitable for analyzing field behavior. T. Takuma and T. Kawamoto [1] reported such numerical results performed by the charge simulation method (CSM) for a variety of configurations with zero contact angles. The contact-point field computed by the boundary element method (BEM) is also presented in Section 3.5 for other simple cases including those with finite volume conduction or surface conduction.

Figure 3.18 shows one such configuration, a rounded conductor slab resting on a dielectric plate with dielectric constant ε_B. The slab (thickness is $2R$) is rounded with radius R. This configuration has three parameters governing the field in both 2D and AS cases: D/R, L/R, and ε_s. The field behavior in Fig. 3.18 considerably differs from that in the above-explained cases shown in Figs. 3.1 and 3.8. First, the contact condition in Fig. 3.18 is neither a single point as in previous AS cases nor a line as in previous 2D cases, but occurs over a surface. Second, the maximum field strength does not appear at the contact point. The latter phenomenon is important, e.g., in a gaseous insulating system, because Paschen's law indicates

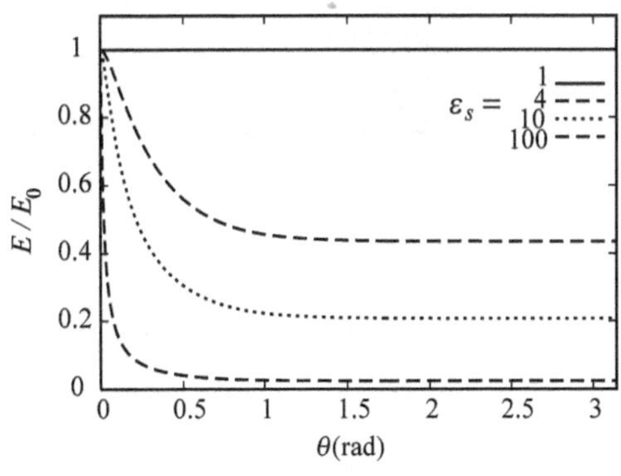

Fig. 3.17 Electric field distribution on the ε_B side along the surface of a dielectric cylinder as shown in Fig. 3.16, where the cylinder is a perfect dielectric [8]. © 2001 IEEE

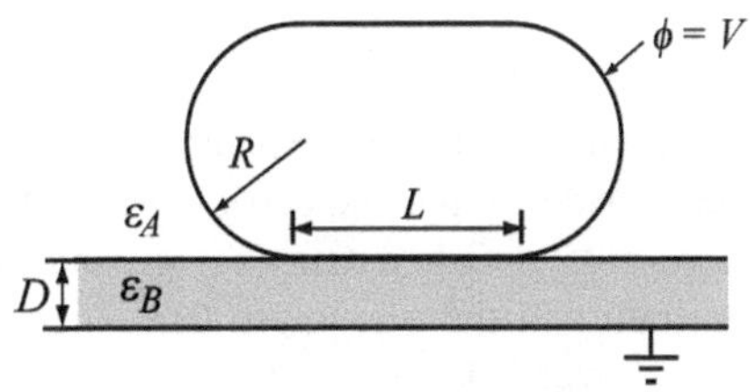

Fig. 3.18 Rounded conductor slab on a solid dielectric plate

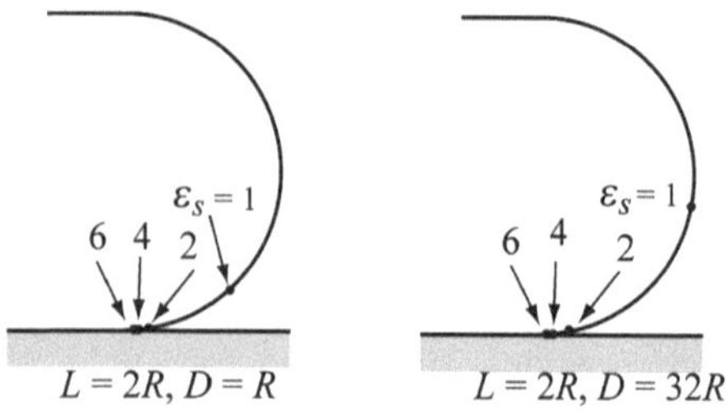

Fig. 3.19 Position of maximum field strength on the conductor surface in AS cases of Fig. 3.18 [1]. © 1984 IEEE

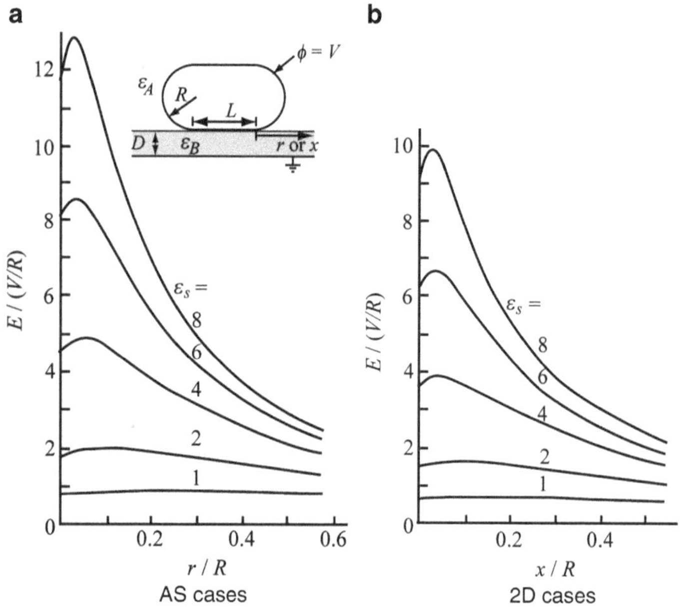

Fig. 3.20 Distribution of the field strength on the ε_A side along the surface of the solid dielectric plate ($D/R = L/R = 2$) [1]. © 1984 IEEE

that the discharge inception voltage is minimal at a certain nonzero product of pressure p and gap length d. The point of maximum field strength is demonstrated in Fig. 3.19 for ε_s values ranging from 1 to 6 in AS cases. It approaches the contact point with increasing ε_s, but does not coincide with it within the computed range of ε_s up to 20.

Figures 3.20a and b show the distributions of field strength normalized by V/R on the ε_A side of the dielectric surface for $D/R = L/R = 2$ for AS and 2D cases, respectively. Other data (not given here) show that both the contact-point field strength and the maximum field strength are almost independent of L/R for L/R values over 2, but rapidly increase with ε_s.

3.4.3 Approximate Expressions of the Contact-Point Field for a Zero Contact Angle

Numerically calculated field strengths E_C (E_{CA} on the ε_A side and E_{CB} on the ε_B side) at the point of contact can be approximated by the following expressions:

$$E_C = f(\varepsilon_s) \cdot E_{C1}, \text{ where}$$

$$f(\varepsilon_s) = \frac{(\varepsilon_s^k + 1)\varepsilon_s}{2} \text{ on the } \varepsilon_A \text{ side, and}$$

$$f(\varepsilon_s) = \frac{\varepsilon_s^k + 1}{2} \text{ on the } \varepsilon_B \text{ side.} \tag{3.11}$$

In these equations, ε_s is $\varepsilon_B/\varepsilon_A$, and E_{C1} is the value for $\varepsilon_s = 1$, i.e., the field strength at the corresponding point without solid dielectric ε_B. Table 3.2 summarizes the values of E_{C1} for a few simple configuration presented later in Section 3.5.2. The constant k, which depends on the configuration, ranges between 0 and 1 in all the cases that were examined for a zero contact angle [1, 2]. As shown in Eqs. 3.1 and 3.2, the limiting case of $k = 1$ in Eq. 3.11 occurs for a conducting sphere or cylinder resting on an infinitely thick solid dielectric with ε_B.

Figure 3.21 presents E_{CB}/E_{C1} (the numerically calculated field strengths on the ε_B side at a contact point normalized by E_{C1}) for ε_s values ranging from 1 to 20 for the two configurations shown on the right side of the figure; the grey areas indicate dielectric solids with dielectric constant ε_B. One configuration is the same as that given in Fig. 3.8 for AS and 2D cases, and the other is a dielectric hemisphere (or semi-cylinder in 2D) placed between a conducting sphere (or cylinder) and a conducting plane. The curves given by Eq. 3.11 are also plotted with suitably chosen values of k. They agree considerably well with the numerical results obtained by the CSM, in particular, for ε_s values up to 10.

Another example is a case of 2D surface contact as shown on the right side of Fig. 3.22, where a solid dielectric ε_B is located between two parallel plane electrodes. The solid edge has the shape of an elliptic quadrant in cross section with major and minor axes a and b. This arrangement is also used in Section 3.5.2 to elucidate the effect of conduction. Figure 3.22 shows the electric field strength on the ε_B side along the solid dielectric surface (interface) for $\varepsilon_s = 4$ as a function of θ for three values of ratio b/a. It is clear that the electric field is more intensified near the contact point with decreasing b/a, probably because the radius of curvature becomes smaller near the contact point. It can also be seen from the figure that the peak (maximum) electric field is not at the contact point and that the location of the peak field gets closer to the contact point with decreasing b/a.

The contact-point field strength numerically calculated by the BEM is compared with that from Eq. 3.11 in Fig. 3.23. The field strength is given as a normalized value on the ε_B side, and the values of constant k used in Eq. 3.11 are specified on the graph. A similar comparison is made for other more practical configurations in Section 5.2.

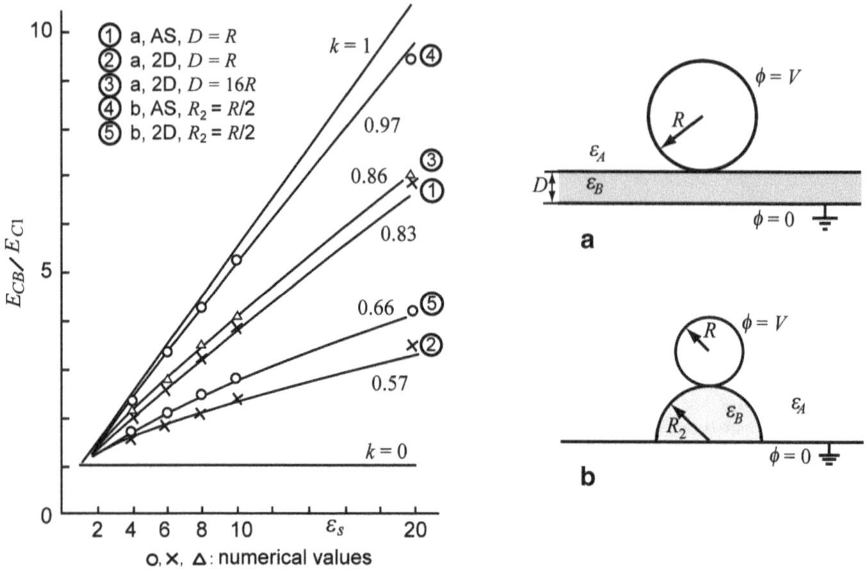

Fig. 3.21 Field strengths for the configurations shown in the figure (**a** and **b**) on the ε_B side at the contact point as a function of ε_s. The symbols represent numerically determined values and the lines were generated using Eq. 3.11 with the k values indicated [1]. © 1984 IEEE

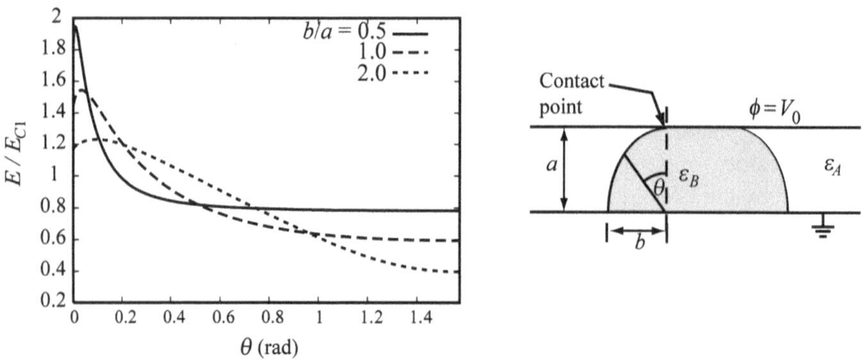

Fig. 3.22 Electric field on the ε_B side along the surface of a 2D solid dielectric in contact with parallel plane electrodes for $\varepsilon_s = 4$ [8]. © 2001 IEEE

3.4.4 Summary of the Contact-Point Field for a Zero Contact Angle

A variety of configurations are possible for the contact of a conductor with a round solid dielectric. Figures 3.24a and b summarize field strength E_{CB} at the point of

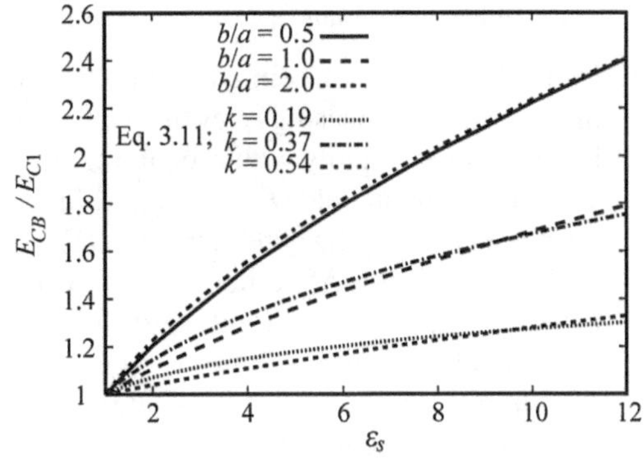

Fig. 3.23 Numerically calculated contact-point field (ε_B side) in comparison with the results from Eq. 3.11 (i.e., the lines associated with a k value) for the configuration shown in Fig. 3.22 [8]. © 2001 IEEE

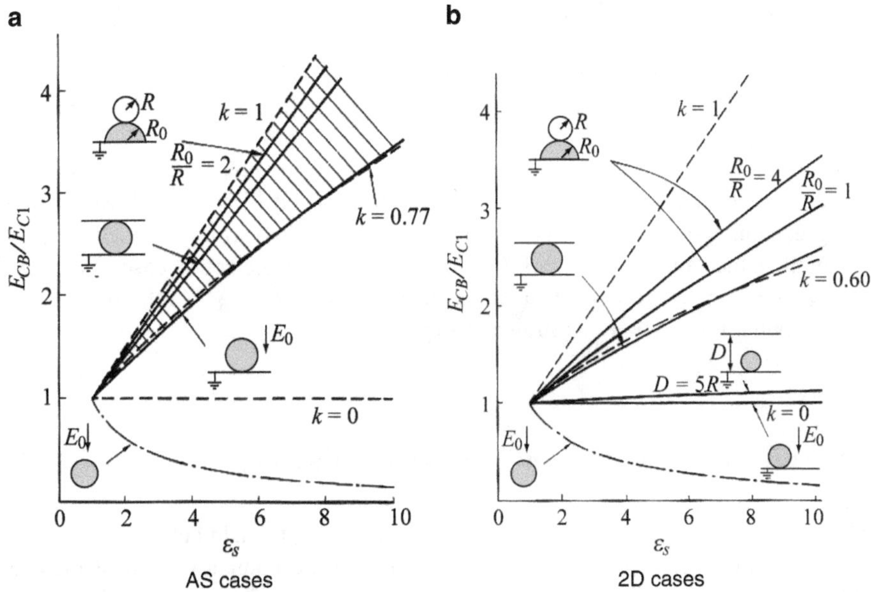

Fig. 3.24 Field strength on the ε_B side at the point of contact with a round dielectric having radius R [2]. © 1991 IEEE

contact on the ε_B side in some simple configurations for AS and 2D cases, respectively. The various configurations are shown in the figures; the shaded areas indicate solid dielectric ε_B. The ordinate E_{CB} is normalized by E_{C1}, as explained in Section 3.4.3. The field strength at the surface of dielectric solid ε_B existing isolated in space under a uniform field is also shown for comparison, although this situation does not represent a contact field and the field strength decreases with increasing ε_s in this case, unlike the case for contact fields. The curves generated by Eq. 3.11 are also plotted for several values of k.

Some of the significant features are summarized from the above results as follows:

1. In all the calculated configurations, the field strength at the point of contact increases with increasing ε_s, even on the ε_B side. This means that the field strength increases more rapidly with ε_s on the ε_A side than it does for a linear relationship.
2. The contact-point field (E_{CA} or E_{CB}) is higher for AS configurations than it is for the corresponding 2D conditions. E_{CB}/E_{C1} on the ε_B side lies in the hatched area of Fig. 3.24a for AS arrangements, which is the area between the curves generated by Eq. 3.11 for $k = 0.77$ and 1. In 2D conditions, however, the corresponding values of k are 0 and 1.
3. The shorter the distance of the contact point from the other (counter) conductor is, the higher the contact field becomes. But this characteristic holds true also for E_{C1} (the value for $\varepsilon_B = \varepsilon_A$).
4. Both the contact field and E_{C1} increase with reducing radius R of the conductor. However, this field enhancement is higher for the contact field than for E_{C1}, thus resulting in an increase of the normalized field strength with R_0/R, where R_0 is the radius of the solid dielectric ε_B, The limiting case of $k = 1$ occurs when R_0 is infinitely large, as explained in Section 4.1.1, or when R is infinitesimally small.
5. The approximate relationship of Eq. 3.11 seems to hold good for the field behavior in a variety of configurations with a round dielectric. However, the agreement is not good when the increase in contact field with ε_s is small, i.e., for small values of k. For small D/R in the sphere-to-dielectric case (Fig. 3.8) explained in Section 3.3.3, a linear characteristic of E_{CB} in relation to ε_s is somewhat more suited, as shown in Fig. 3.15.

3.5 Effect of Volume and Surface Conduction

This section discusses the effect of finite conduction on the field behavior in configurations with a zero contact angle, as analyzed by the boundary element method (BEM) [8, 9]. We again confine our interest to a 50-Hz power frequency in this section.

3.5.1 Solid Dielectric Cylinder with Volume Conduction Under a Uniform Field

The effect of volume conduction on the field behavior is studied for the 2D configuration of a solid dielectric cylinder with ε_B lying on a plane conductor under a uniform electric field. This configuration corresponds to Fig. 3.16 for a perfect dielectric without conduction, but here the solid has finite conductivities σ and σ_s, as shown in Fig. 3.25. An analytical solution can be obtained by applying an infinite number of dipole images in a similar form to that in the case of a perfect

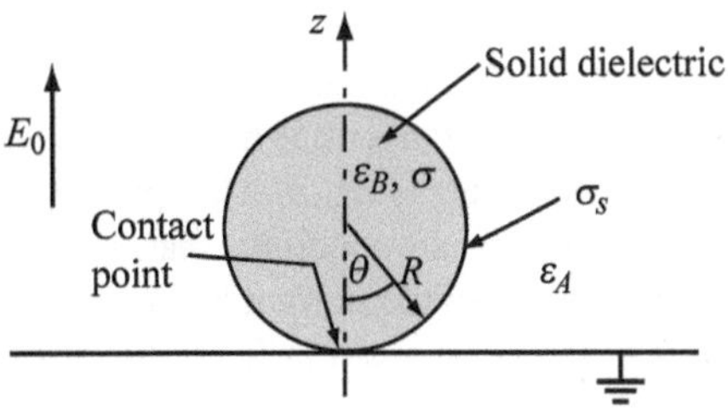

Fig. 3.25 A solid dielectric cylinder with volume and surface conductivity lying on a grounded conductor plane under an external uniform electric field

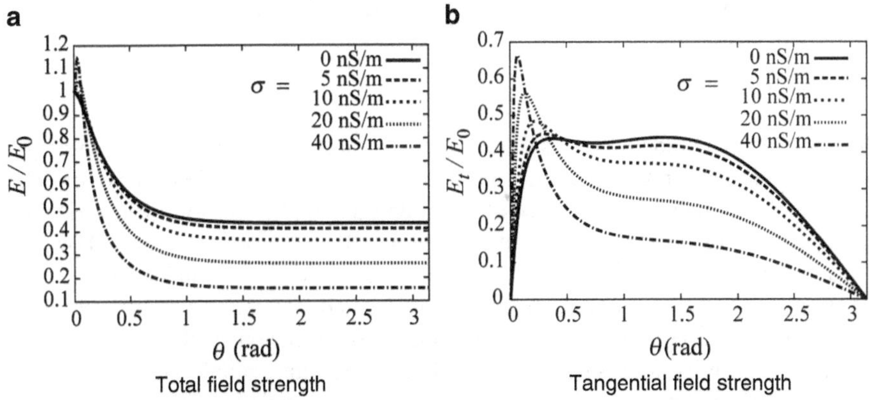

Fig. 3.26 Electric field distribution on the ε_B side along the surface of a dielectric cylinder ($\varepsilon_A = 1$ and $\varepsilon_B = 4$) with volume conduction under a uniform field [8]. © 2001 IEEE

dielectric, but replacing the dielectric constant (relative permittivity) ε with a complex expression:

$$\dot{\varepsilon}_B = \varepsilon_B - j\varepsilon_B' = \varepsilon_B + \frac{\sigma}{j\omega\varepsilon_0} = \varepsilon_B - j\frac{\sigma}{\omega\varepsilon_0} \tag{3.12}$$

Figure 3.26 presents the normalized electric field on the ε_B side along the surface of a cylinder (ε_A =1 and ε_B =4) with volume conductivity σ. As was also the case in Fig. 3.17 for a solid dielectric without conduction, the normalized field strength E_{CB}/E_0 at the contact point is equal to unity, irrespective of the value of σ. Another interesting characteristic is that the position of the maximum electric field shifts from the contact point with the presence of volume conduction, although the maximum is always at the contact point for the case of zero conductivity, as shown in Fig. 3.17. The shift of the maximum field is noticeable in Fig. 3.26a when σ is higher than about 10 nS/m. This value corresponds to $\varepsilon'_B = 3.6$ in Eq. 3.12 for 50 Hz, which is comparable to the value of ε_B.

The tangential field shown in Fig. 3.26b must be zero at the contact point because of the symmetrical characteristic of the field for a zero contact angle. The maximum tangential field becomes higher and is located closer to the contact point with increasing σ. Comparing the total field and the tangential field in Fig. 3.26, we can conclude that, because the peak positions of E and E_t are not coincident, both the normal and tangential components of the field contribute to the shift of the peak position from the contact point.

3.5.2 *Other Configurations with Volume Conduction*

3.5.2.1 Point and Line Contact

We here consider some other simple configurations with a zero contact angle for which numerical approaches must be applied to analyze the field behavior. The configurations include the AS case of Fig. 3.25 (i.e., a spherical solid dielectric on a plane conductor), and both 2D and AS cases of the two configurations shown in Fig. 3.27. In Fig. 3.27a, a solid dielectric hemisphere (or semi-cylinder) of dielectric constant ε_B and volume conductivity σ is located between two parallel plane conductors. This configuration can be seen as an extension of those in Fig. 3.16 or 3.25 because its field is equivalent to that of an infinite number of spherical (or cylindrical) dielectrics existing in series under a field equal to V_0/R_0. In Fig. 3.27b and Fig. 3.21b, in contrast, the configuration of a solid dielectric (with ε_B and σ)

Table 3.2 Values of the contact-point field E_{C1} for $\varepsilon_B = \varepsilon_A$

Figure showing the configuration	E_{C1}	
	AS	2D
Figure 3.25	E_0	E_0
Figure 3.27a	V_0/R_0	V_0/R_0
Figure 3.27b	1.770 V_0/R	1.315 V_0/R

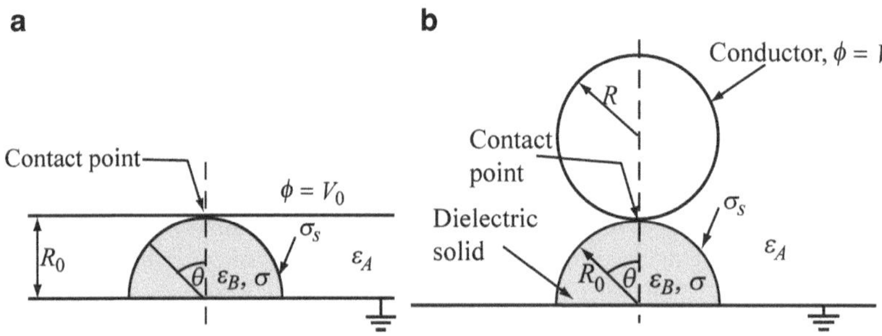

Fig. 3.27 A dielectric hemisphere (or semi-cylinder) between (**a**) two plane conductors or (**b**) a conducting sphere (or cylinder) and a plane conductor

between a conducting sphere (or cylinder) and a plane conductor is examined to elucidate the field behavior under a nonuniform external field. The radius R is taken to be the same as R_0 of the dielectric ε_B in this section. We again use the contact-point electric field E_{C1} for $\varepsilon_B = \varepsilon_A$ to normalize the resulting field. Table 3.2 summarizes the value of E_{C1} for each configuration given in Figs. 3.25 and 3.27.

The contact-point electric field E_{CB} on the ε_B side for these configurations is presented in Fig. 3.28. The values for a perfect dielectric (zero conductivity) are given with lines and those with volume conduction for $\varepsilon_B/\varepsilon_A = 2$ or 4 with the respective symbols. The horizontal axis is the absolute value $|\dot{\varepsilon}_B|$, which, as shown in Eq. 3.12, depends on ε_B and σ. For $\sigma = 0$, then, $|\dot{\varepsilon}_B| = \varepsilon_B$. It is clear that E_{CB} in all cases of Fig. 3.27 is enhanced by the increase of $|\dot{\varepsilon}_B|$, either as a result of an increase

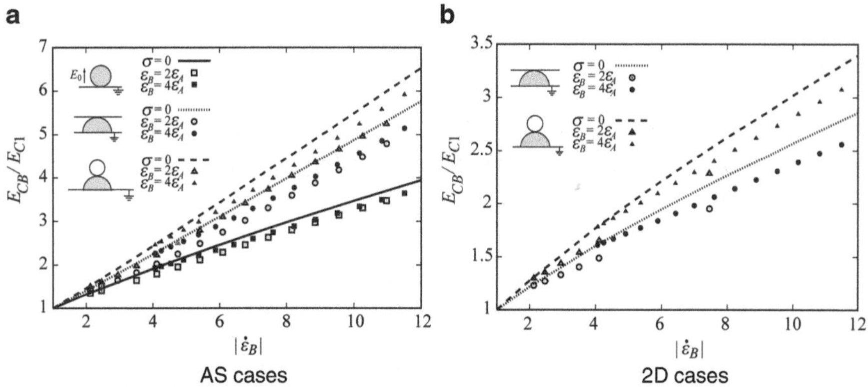

Fig. 3.28 Contact-point electric field on the ε_B side as a function of the absolute value of the complex dielectric constant. The field values are given as lines (no volume conduction) and symbols (with volume conduction) [8]. © 2001 IEEE

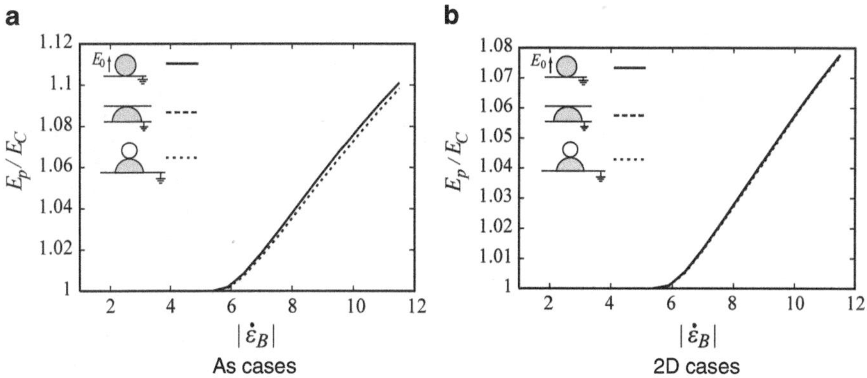

Fig. 3.29 Ratio of peak electric field to contact-point field on the ε_B side as a function of the absolute value of the complex dielectric constant [8]. © 2001 IEEE

in ε_B (if $\sigma = 0$) or an increase in σ for fixed ε_B. However, the contact-point field is highest in the perfect dielectric case for the same value of $|\dot{\varepsilon}_B|$. Comparing Figs. 3.28a and b, we can see that the AS case (point contact) gives a higher E_{CB} than the 2D case (line contact) for the same configuration and the same value of $|\dot{\varepsilon}_B|$.

As already mentioned, the electric field on the dielectric interface may not peak at the contact point due to the effect of volume conduction. Figure 3.29 shows the ratio of the peak field E_p to the contact-point field E_C on the ε_B side for $\varepsilon_A = 1$ and $\varepsilon_B = 4$ when σ is varied. $E_p/E_C > 1$ indicates the shift of the peak-field position from the contact point. From the figure, it is clear that the ratio E_p/E_C hardly varies among the configurations concerned. The shift of the peak-field position is noticeable in all the cases studied here when the imaginary part $\sigma/\omega\varepsilon_0$ is comparable to or larger than ε_B.

3.5.2.2 Surface Contact

Another example is the 2D configuration of a curved solid dielectric with surface contact shown in Fig. 3.30. The field behavior for a perfect dielectric of zero

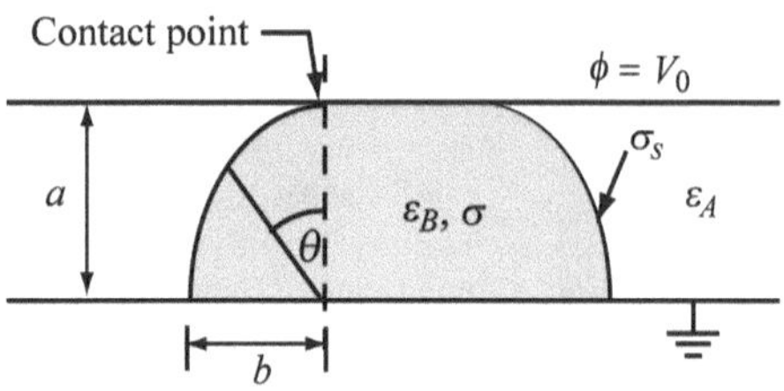

Fig. 3.30 Curved solid dielectric having its surface in contact with two parallel plane electrodes

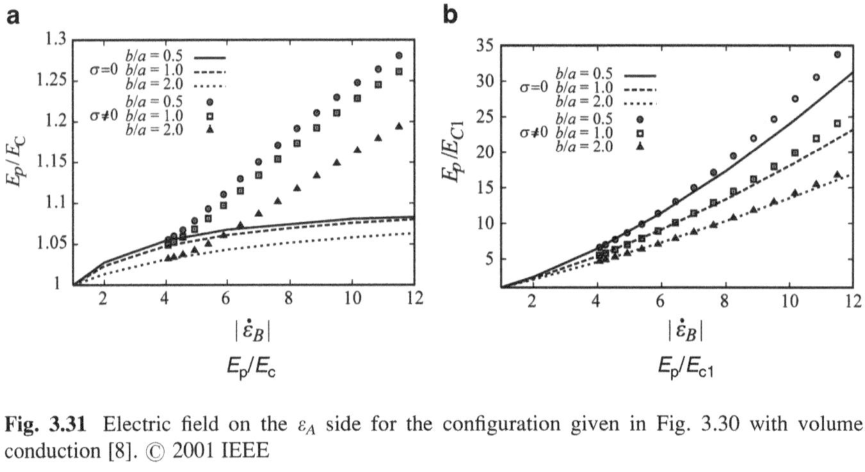

Fig. 3.31 Electric field on the ε_A side for the configuration given in Fig. 3.30 with volume conduction [8]. © 2001 IEEE

conductivity is given in Figs. 3.22 and 3.23 in Section 3.4.3. The peak electric field does not appear at the contact point, even for the case of zero conductivity.

The effect of volume conduction in the solid dielectric is presented in Fig. 3.31, which shows the ratios E_p/E_C and E_p/E_{C1} on the ε_A side for $\varepsilon_A = 1$ and $\varepsilon_B = 4$. In this figure, E_p is the peak electric field, E_C is the contact-point field, and E_{C1} is the value for ε_s $(= \varepsilon_B/\varepsilon_A) = 1$, i.e., the field strength at the corresponding point without dielectric ε_B. In Fig. 3.31a, E_p is always larger than E_C, which indicates that the position of the peak field is shifted from the contact point both with and without conductivity. Figure 3.31b shows that the peak electric field becomes higher with increasing volume conductivity σ, compared to the solid lines for zero conductivity at the same value of $|\dot{\varepsilon}_B|$. This behavior is in contrast with that in the configurations with point or line contact, in which the field is lower with higher σ at the same value of $|\dot{\varepsilon}_B|$.

3.5.3 *Effect of Surface Conduction*

The field behavior in the presence of surface conduction (finite σ_s with zero σ) is examined for an exemplary configuration of Fig. 3.27a with $\varepsilon_A = 1$ and $\varepsilon_B = 4$. Recalling the boundary condition of the electric field on the dielectric interface with surface conduction, which is given as Eq. 1.21,

$$\dot{\varepsilon}_B \dot{E}_{nB} - \dot{\varepsilon}_A \dot{E}_{nA} = \frac{1}{j\omega\varepsilon_0} \nabla_s \cdot \dot{\mathbf{J}}_s = \frac{1}{j\omega\varepsilon_0} \nabla_s \cdot \sigma_s \dot{\mathbf{E}}_t, \tag{3.13}$$

we can expect that the influence of σ_s would be marked when the tangential field E_t changes rapidly with distance. For this reason, the field behavior near a contact point with a zero contact angle is affected even by small levels of surface conduction as E_t steeply changes near the contact point (see the case of zero conductivity in Fig. 3.26b, for example).

3.5.3.1 Lower Conductivity

The electric field near the contact point in the 2D case of Fig. 3.27a is presented in Fig. 3.32 for $\sigma_s < 1$ nS as a function of angle θ (in radians) from the contact point. Angle θ is indicated in Fig. 3.27a. The shift of the position of the peak electric field on the ε_A side from the contact point can be observed at σ_s values as small as 0.8 nS in Fig. 3.32a, whereas the field hardly changes with σ_s on the ε_B side in Fig. 3.32b. In contrast, the tangential field component (not shown here) hardly changes with σ_s in this range, implying that the shift of the position of the peak field is mainly due to the normal component of the electric field.

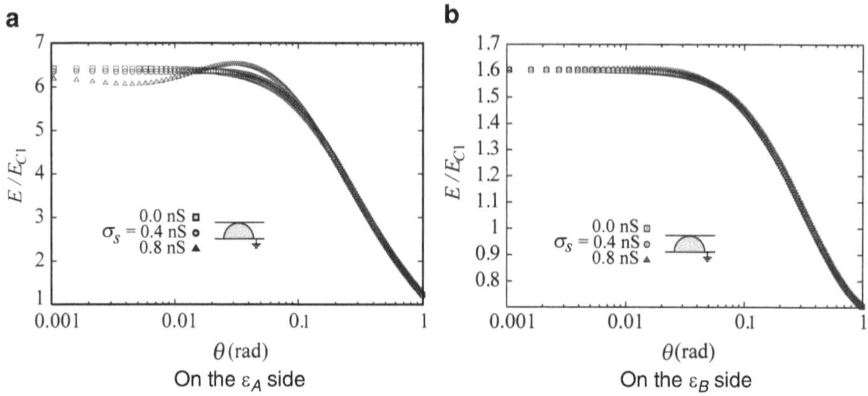

Fig. 3.32 Electric field near the contact point in the 2D case of the configuration shown in Fig. 3.27a [9]. © 2002 IEEE

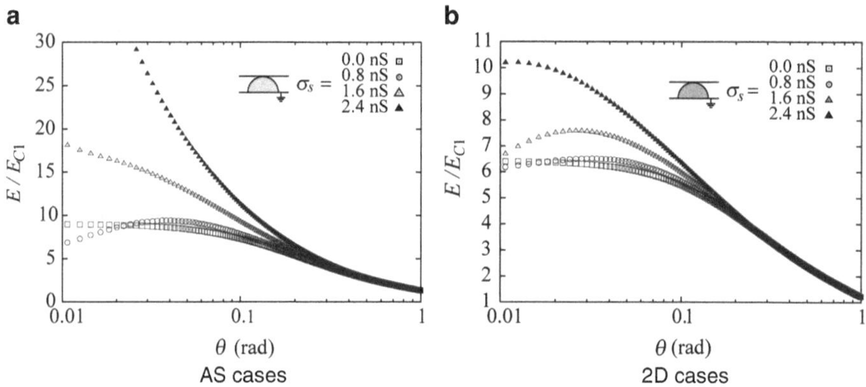

Fig. 3.33 Electric field distribution on the ε_A side along the dielectric interface [9]. © 2002 IEEE

3.5.3.2 Higher Conductivity

For higher values of surface conduction, the electric field on the ε_A side in the AS and 2D cases of Fig. 3.27a is given in Figs. 3.33a and b, respectively. These figures show that the effect of σ_s becomes noticeable, i.e., the peak field is not located at the contact point, at $\sigma_s = 0.8$ nS. With a further increase in σ_s, the conductivity considerably enhances the electric field near the contact point, and the position of the peak field moves closer to the contact point.

Figure 3.34 summarizes the normalized peak field strength on the ε_A side as a function of σ_s for the two configurations given in Figs. 3.27a and b. The effect of σ_s is more pronounced for the Fig. 3.27b configuration, in which the field is

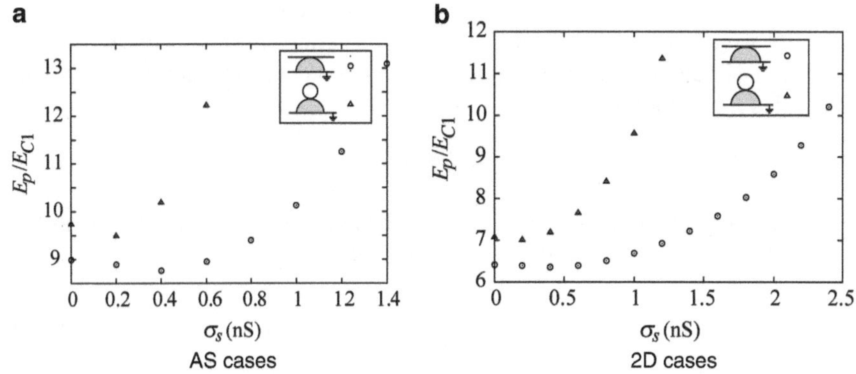

Fig. 3.34 Normalized peak electric field on the ε_A side along the dielectric interface [9]. © 2002 IEEE

Fig. 3.35 Normalized peak electric field on the ε_A side for the 2D case of Fig. 3.30 with surface contact [9]. © 2002 IEEE

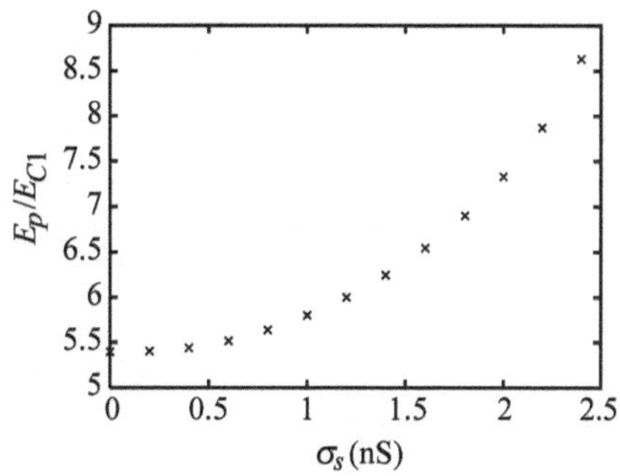

basically more nonuniform, than for that in Fig. 3.27a. The relative variations of the peak field with σ_s are similar for the AS and 2D cases.

Figure 3.35 shows the peak electric field E_p on the ε_A side versus σ_s for a configuration with surface contact, i.e., for the 2D case of Fig. 3.30 with $b/a = 1$, $\varepsilon_A = 1$, and $\dot{\varepsilon}_B = \varepsilon_B = 4$ (no volume conduction). The variation of E_p basically resembles that for the line contact given in Fig. 3.34b. However, the normalized field magnitude is smaller for the configuration with surface contact, as shown in Fig. 3.35.

3.5.4 *Approximate Treatment for Surface Conduction*

3.5.4.1 Limiting Field Distribution for High Conductivity

As explained in Section 2.3.4, high surface conductivity governs the potential distribution along a solid dielectric surface (interface) and thus creates a constant field distribution there in the case of constant surface conductivity. In the 2D case of Fig. 3.36, which is inverted from Fig. 3.27a for easier viewing, a linear potential distribution should be established in relation to θ along the arc-like round interface with radius R as follows:

$$\phi = 2\,\theta\,V_0/\pi, \tag{3.14}$$

where V_0 is the potential difference between the two parallel plane electrodes. This leads to the following tangential field component E_t:

$$E_t = \frac{\partial \phi}{\partial (R\theta)} = \frac{2V_0}{\pi R} = 0.637V_0/R. \tag{3.15}$$

The normal component E_n is approximately given near the contact point as

$$E_n = \frac{\phi}{l_n} = \frac{2\theta V_0}{\pi} \cdot \frac{\cos\theta}{R(1-\cos\theta)}. \tag{3.16}$$

where l_n is the same as line $\overline{\mathrm{ab}}$ in Fig. 3.8, i.e., the distance from (and normal to) the dielectric surface to the lower electrode (see Fig. 3.36). In the close vicinity of the contact point, Eq. 3.16 furthermore becomes

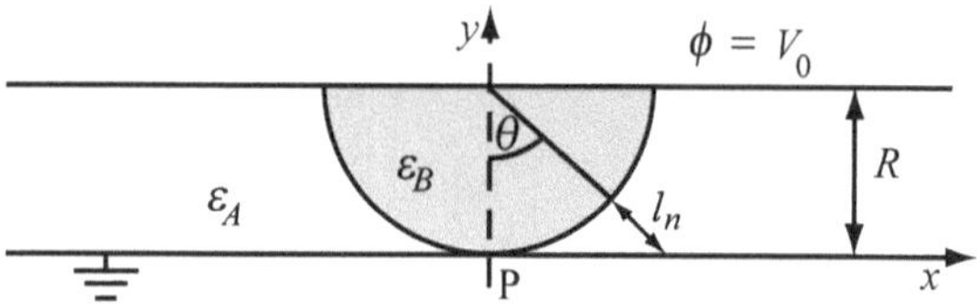

Fig. 3.36 2D line contact of a semi-cylindrical dielectric with surface conduction, similar to the arrangement given in Fig. 3.27a

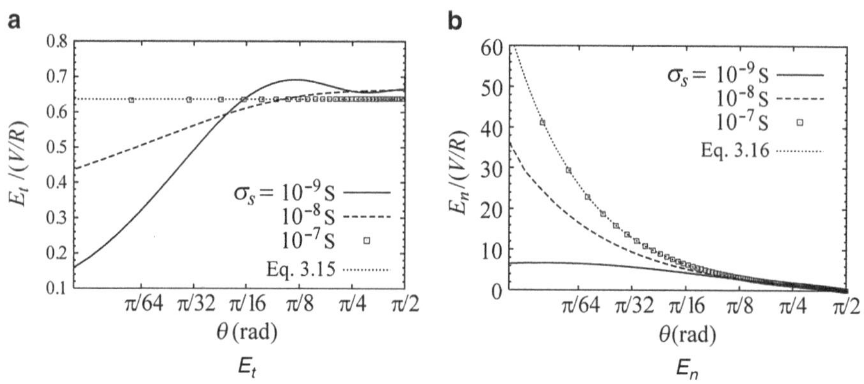

Fig. 3.37 Tangential and normal components of the electric field on the ε_A side for high surface conductivity for the configuration shown in Fig. 3.36

$$E_n = 4V_0/(\pi R\theta) = (1.27/\theta)V_0/R. \tag{3.17}$$

These expressions mean that for sufficiently high surface conductivity σ_s, the field strength near the contact point is independent of σ_s and increases as the reciprocal of θ when the contact point is approached. Figures 3.37a and b show E_t and E_n for some values of σ_s for the configuration given in Fig. 3.36 (2D case of Fig. 3.27a), together with the values from Eqs. 3.15 and 3.16. The horizontal axis is enlarged in a logarithmic scale near the contact point. It can be seen that at $\sigma_s = 10^{-7}$ S, E_t and E_n coincide with the limiting values for high conductivity.

Similar quantitative estimation may be carried out for the case of point contact in an AS arrangement of Fig. 3.27a. However, significant differences from the 2D line contact case are found in the AS configuration, even for high, constant σ_s. First, the field strength for all values of θ depends on σ_s, and second, a constant field distribution is not established along the dielectric interface. This is because the resistive (or conductive) impedance is infinitely high at the contact point due to the infinitesimally narrow current passage there. It is anticipated that the field strength near the contact point will increase more rapidly with increasing σ_s in the AS point contact case than in the 2D line contact case, which is demonstrated above in Fig. 3.33.

3.5.4.2 Approximate Evaluation of the Effect of Surface Conduction

In a similar approach to that used in Section 2.3.5, the resistive (or conductive) impedance Z_R is compared with the capacitive impedance Z_C in a unit length vertical to the figure space. For the range of x from 0 to R in Fig. 3.36,

$$Z_R = \frac{\pi R}{2\sigma_s}, \text{ and } Z_C = \frac{1}{\omega C} = \frac{D}{2\pi f \varepsilon_E \varepsilon_0 R}, \tag{3.18}$$

where D is the electrode separation, equal to R, and the equivalent dielectric constant ε_E is approximated with ε_B in this case. Thus,

$$\frac{Z_R}{Z_C} = \frac{\pi^2 f \varepsilon_B \varepsilon_0 R}{\sigma_s}. \tag{3.19}$$

Considering that a 10% change in impedance corresponds to $Z_R/Z_C = 2$, which is explained in Section 2.3.5, the effect of surface conduction becomes significant at the following value of σ_s:

$$\sigma_s = \pi^2 f \varepsilon_B \varepsilon_0 R/2 = 8.7R \text{ nS}, \tag{3.20}$$

for $f = 50$ Hz, $\varepsilon_A = 1$ and $\varepsilon_B = 4$. If we consider the close vicinity of the contact point instead of the horizontal length of $2R$, the impedances are approximated as

$$Z_R = \frac{\theta R}{\sigma_s}, \text{ and } Z_C = \frac{D_y}{2\pi f \varepsilon_E \varepsilon_0 R \sin\theta}, \; D_y = R - R\cos\theta, \tag{3.21}$$

where D_y is the distance in the y-direction, This leads to $Z_R/Z_C = 4\pi f \varepsilon_0 \varepsilon_B R/\sigma_s$, and the effective value of σ_s is derived as:

$$\sigma_s = 2\pi f \varepsilon_B \varepsilon_0 R = 11.1R \text{ nS}, \tag{3.22}$$

for the conditions mentioned above. However, these values are about one order of magnitude larger than the numerically calculated results in which the effect of surface conduction is noticeable at about $\sigma_s = 0.8$ nS for $R = 1$ m.

References

1. Takuma, T., Kawamoto, T.: Field intensification near various points of contact with a zero contact angle between a solid dielectric and an electrode. IEEE Trans. Power. Appar. Syst. **103**(9), 2486–2494 (1984), and Takuma, T., Kawamoto, T.: Field behavior at a contact point of a round electrode with solid dielectric. Proc. 4th ISH (Int. Symp. on High Volt. Eng.):12.04 (1983)
2. Takuma, T.: Field behavior at a triple junction in composite dielectric arrangements. IEEE Trans. Electr. Insul. **26**(3), 500–509 (1991), and Takuma, T., Kawamoto, T.: Electric field at various points of contact between rounded dielectric and electrode. Proc. 5th ISH:33.09 (1987)
3. Wensley, J.H., Parker, F.W.: The solution of the electric field problems using a digital computer. Electr. Energy **1**, 12–16 (1956)
4. Kao, K.C., Harker, T.: The calculation of the electric field for an infinite dielectric plate between two spherical electrodes. Proc. IEE Part C **109**(16), 293–298 (1962)
5. Binns, D.F., Randall, T.J.: Calculation of potential gradients for a dielectric slab placed between a sphere and a plane. Proc. IEE Part C **114**(10), 1521–1528 (1967)
6. Chow, Y.L., Srivastava, K.D., Charalambous, C.: Electrostatic field between a conducting sphere and a dielectric-coated electrode. J. Electrost. **11**(3), 167–178 (1982)
7. Weiss, P.: Rotationssymmetrische Zweistoffdielektrika. Diss. Tech. Univ. Munich (in German) (1972)
8. Techaumnat, B., Hamada, S., Takuma, T.: Electric field behavior near a contact point in the presence of volume conductivity. IEEE Trans. Dielectr. Electr. Insul. **8**(6), 930–935 (2001)
9. Techaumnat, B., Hamada, S., Takuma, T.: Electric field behavior near a zero-angle contact point in the presence of surface conductivity. IEEE Trans. Dielectr. Electr. Insul. **9**(4), 537–543 (2002)
10. Techaumnat, B., Takuma, T.: Electric field and force on a conducting sphere in contact with a dielectric solid. J. Electrost. **64**(3–4), 165–175 (2006)
11. Takuma, T., Kawamoto, T.: Field behavior when a solid dielectric is in contact with an electrode. Conf. Rep. Tech. Comm. on Electrical Discharge, IEEJ (Inst. Electr. Eng. Japan): ED-83-14 and ED-83–50 (1983) (in Japanese)

Chapter 4
Electric Field Behavior for the Common Contact of Three Dielectrics

Introduction

When the interfaces of three dielectrics that meet at a point are straight or linear in cross section, the electric potential and the field strength are proportional to l^n and l^m, respectively, near the point of contact, where l is the distance from the contact point and $m = n - 1$. We can derive an equation for determining exponent n in a similar way to that used for two dielectrics. As far as we know, the analytical study on the field behavior for this configuration has been reported only recently in a few papers by T. Takuma et al. [1, 2]. They have also reported various applications of the derived equation for determining n, and the effect when one of the interface edges is curved near the contact point; the latter analysis was carried out using a numerical field calculation method [3]. More complex configurations with three dielectrics, such as a solid dielectric sphere in contact with another solid sphere, are dealt with in later chapters.

4.1 Contact of Straight Dielectric Interfaces

4.1.1 Basic Field Behavior

Figure 4.1 shows the basic configuration in which three dielectrics with dielectric constants (relative permittivities) ε_A, ε_B, and ε_C meet at a central point P. In this chapter, as in previous chapters, a vacuum can be regarded as one of the dielectrics having absolute permittivity ε_0 (electric constant) or a unit dielectric constant. Thus, a vacuum is equivalent to a gaseous dielectric from the standpoint of the electrostatic field. We can treat the configuration as two-dimensional (2D) in the close neighborhood of the contact point (interface edge) where all the interfaces (or solid dielectric surfaces) are straight or linear in cross section. In this situation, we express the electric potential in the 2D polar coordinates (r, θ), as represented

T. Takuma and B. Techaumnat, *Electric Fields in Composite Dielectrics and their Applications*, Power Systems, DOI 10.1007/978-90-481-9392-9_4,

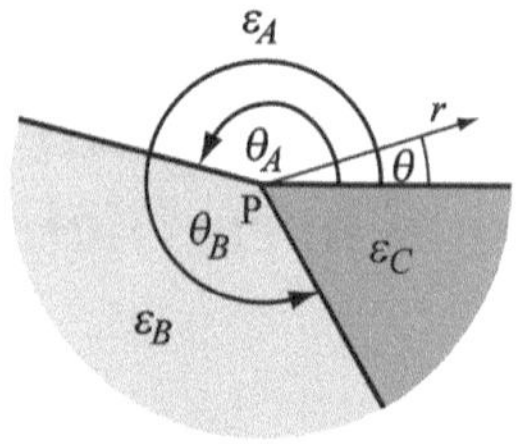

Fig. 4.1 Contact of three straight dielectric interfaces

in Fig. 4.1. The origin of the coordinates is taken at the contact point. As in the two-dielectric cases explained in Chapter 2, the field behavior can be analyzed by applying the variable-separation method to Laplace's equation in the 2D polar coordinates. The calculation procedure is explained in Section 8.1.5, and only the essential results are given here.

The potentials in the three media in Fig. 4.1 are expressed as a power function of r, i.e., r^n, and the electric fields as r^{n-1}. The exponent n is the smallest positive solution of the following equation, which is derived from Eq. 8.13, the determinant of the coefficient matrix in Section 8.1.5,

$$\begin{aligned}&(\varepsilon_A^2+\varepsilon_B^2)\varepsilon_C\sin A\sin(A-B)\cos(B-C)+(\varepsilon_A^2+\varepsilon_C^2)\varepsilon_B\sin A\cos(A-B)\sin(B-C)\\&-(\varepsilon_B^2+\varepsilon_C^2)\varepsilon_A\cos A\sin(A-B)\sin(B-C)-2\varepsilon_A\varepsilon_B\varepsilon_C[1-\cos A\cos(A-B)\cos(B-C)]=0\end{aligned} \tag{4.1}$$

In this equation, $A = n\theta_A$, $B = n\theta_B$, and $C = 2\pi n$, and angles θ_A and θ_B are given in the figure. Another equation can also be derived from the matrix by applying the Mathematica as follows:

$$\begin{aligned}8\varepsilon_A\varepsilon_B\varepsilon_C &- (\varepsilon_A+\varepsilon_B)(\varepsilon_A+\varepsilon_C)(\varepsilon_B+\varepsilon_C)\cos C\\&+(\varepsilon_A-\varepsilon_B)(\varepsilon_A-\varepsilon_C)(\varepsilon_B+\varepsilon_C)\cos(2A-C)\\&+(\varepsilon_A+\varepsilon_B)(\varepsilon_A-\varepsilon_C)(\varepsilon_B-\varepsilon_C)\cos(2B-C)\\&-(\varepsilon_A-\varepsilon_B)(\varepsilon_A+\varepsilon_C)(\varepsilon_B-\varepsilon_C)\cos(2A-2B+C)=0\end{aligned} \tag{4.2}$$

This equation, a linear combination of cosine functions, is seemingly quite different from Eq. 4.1, a cubic equation of trigonometric functions, but they are in fact equivalent.

4.1.2 *Applications of the Equations for n*

Equations 4.1 and 4.2 are generally applicable to various configurations composed of three cross-sectionally straight interfaces. One of the three dielectrics may be a vacuum, as mentioned above, and, furthermore, a conductor can be treated as having an infinitely high dielectric constant. As a self-evident result, the equations should also include two-dielectric cases, which correspond to cases with $\theta_B = 2\pi$ in Fig. 4.1. Some of the applications where simple analytical expressions are available are briefly reviewed below.

4.1.2.1 Two Dielectrics Without a Conductor

This configuration corresponds to Fig. 2.4 in Section 2.1.3. Taking $\theta_B = 2\pi$ and $\varepsilon_B = \varepsilon_C$ in Eq. 4.1 or 4.2 leads to Eq. 2.6 or 2.7, as explained there.

4.1.2.2 Two Dielectrics in Contact with a Conductor

Various cases may exist in this category concerned with high-voltage insulation and vacuum technology. Some typical examples are:

- Straight interface of two dielectrics in contact with a plane conductor.
  This configuration is the same as Fig. 2.1, the field behavior of which is explained in detail in Section 2.1.
- Metal edge on a plane solid dielectric.
- Contact of a solid dielectric and a conductor both having a sharp corner.

In each of these cases, a transcendental equation for determining exponent n can be derived from Eq. 4.1 or 4.2 by making one of the three dielectric constants infinitely high. Some of these derivations are explained in Section 5.4.

4.1.2.3 Three Dielectrics with the Same Angle

When the three dielectrics form a symmetrical contact with the same cross-sectional angle of 2π/3, as shown in Fig. 4.2, $\theta_A = 2\pi/3$ and $\theta_B = 4\pi/3$ in Fig. 4.1. Under these conditions, Eq. 4.1 gives

$$\begin{aligned}&\left[(\varepsilon_A^2+\varepsilon_B^2)\varepsilon_C+(\varepsilon_A^2+\varepsilon_C^2)\varepsilon_B+(\varepsilon_B^2+\varepsilon_C^2)\varepsilon_A\right]\sin^2\!A\cos A\\&\quad+2\varepsilon_A\varepsilon_B\varepsilon_C\left(1-\cos^3\!A\right)=0,\end{aligned} \tag{4.3}$$

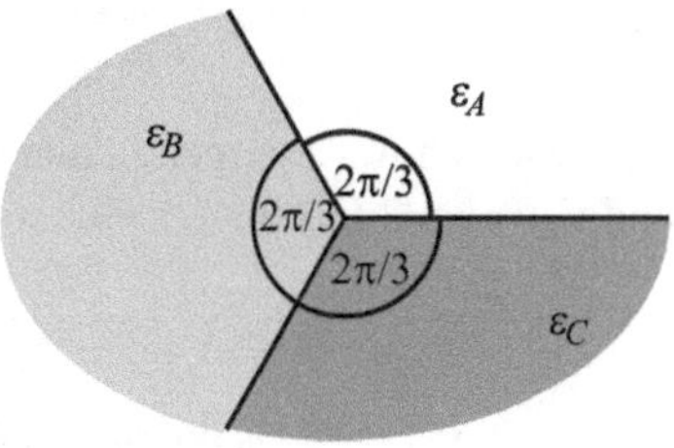

Fig. 4.2 Configuration consisting of three dielectrics with the same edge angle

where $A = 2n\pi/3$. Rearranging this, the following explicit expression giving n can be obtained:

$$\cos A = -\frac{1}{2} \pm \sqrt{\frac{1}{4} - \frac{X}{Y}}, \tag{4.4}$$

where $X = 2\varepsilon_A \varepsilon_B \varepsilon_C$, and

$$\begin{aligned} Y &= (\varepsilon_A^2 + \varepsilon_B^2)\varepsilon_C + (\varepsilon_A^2 + \varepsilon_C^2)\varepsilon_B + (\varepsilon_B^2 + \varepsilon_C^2)\varepsilon_A + 2\varepsilon_A \varepsilon_B \varepsilon_C \\ &= (\varepsilon_A + \varepsilon_B)(\varepsilon_A + \varepsilon_C)(\varepsilon_B + \varepsilon_C). \end{aligned}$$

Thus,

$$\frac{1}{4} - \frac{X}{Y} = \frac{\varepsilon_A(\varepsilon_B - \varepsilon_C)^2 + \varepsilon_B(\varepsilon_A - \varepsilon_C)^2 + \varepsilon_C(\varepsilon_A - \varepsilon_B)^2}{4(\varepsilon_A + \varepsilon_B)(\varepsilon_A + \varepsilon_C)(\varepsilon_B + \varepsilon_C)}. \tag{4.5}$$

Needless to say, Eq. 4.4 constitutes an equivalent expression for any of the three dielectrics, ε_A, ε_B, and ε_C. Equation 4.5 is always positive, its value lying between 0 and 1/4. From this characteristic, we can evaluate exponent n as well as the field behavior. If we consider the lowest positive value of n, we should take the positive sign in Eq. 4.4. Thus, values of 0 and 1/4 for Eq. 4.5 correspond to $A = \pi/2$ and $2\pi/3$, which result in $n = 3/4$ and 1, respectively. The case of no field singularity (not resulting in an infinitely high field) corresponding to $n = 1$ occurs only for the self-evident case of $\varepsilon_A = \varepsilon_B = \varepsilon_C$. The lowest value of n is 3/4. This means the strongest singularity appears when $X \ll Y$, corresponding to the case where one of the dielectrics is a wedge-like conductor with an angle of $2\pi/3$.

4.2 Perpendicular Contact of a Solid Dielectric with Another Solid

Because this case is important as it is encountered in practical configurations in which a solid dielectric supports another solid in a gaseous or liquid dielectric, or a vacuum, the field behavior is analyzed below in more detail.

4.2.1 Equation for Determining n

Figure 4.3 shows a perpendicular support, which corresponds to the case of $\theta_A = \pi/2$ and $\theta_B = 3\pi/2$ in Fig. 4.1. Substituting these values in Eq. 4.1 gives

$$\begin{aligned} &(\varepsilon_B^2 + \varepsilon_C^2)\varepsilon_A \sin^2(2A)/2 + (\varepsilon_A^2 + \varepsilon_B^2)\varepsilon_C \sin^2(2A)/2 \\ &\quad + (\varepsilon_A^2 + \varepsilon_C^2)\varepsilon_B \sin^2 A \cos(2A) + 2\varepsilon_A \varepsilon_B \varepsilon_C \{1 - \cos(2A)\cos^2 A\} = 0. \end{aligned} \tag{4.6}$$

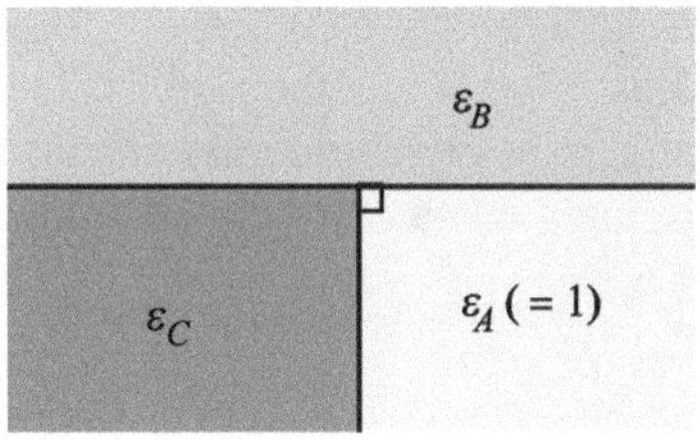

Fig. 4.3 A perpendicular solid dielectric supporting another plane dielectric

In this equation, $A = n\pi/2$. Rearranging this equation, the following simple expression can be obtained [1, 2]:

$$\cos(2A) = -\frac{\varepsilon_A(\varepsilon_B+\varepsilon_C)^2+\varepsilon_C(\varepsilon_A+\varepsilon_B)^2}{\varepsilon_A(\varepsilon_B+\varepsilon_C)^2+\varepsilon_C(\varepsilon_A+\varepsilon_B)^2+\varepsilon_B(\varepsilon_A-\varepsilon_C)^2}$$

or

$$\cos^2 A = \cos^2(n\pi/2) = \frac{\varepsilon_B(\varepsilon_A-\varepsilon_C)^2}{2(\varepsilon_A+\varepsilon_B)(\varepsilon_A+\varepsilon_C)(\varepsilon_B+\varepsilon_C)}. \tag{4.7}$$

Equation 4.7 means that the right-hand side term becomes 0 either when ε_B is infinitely high or when $\varepsilon_C = \varepsilon_A$. The former case corresponds to the solid ε_B being a conductor. In both cases, the solution leads to $n = 1$, and it is self-evident that no field singularity occurs in these cases.

4.2.2 Applications of Eq. 4.7

The right-hand side of Eq. 4.7 is further studied with the following function F of ε_B and ε_C, mainly for the case of the third dielectric being a gas or a vacuum ($\varepsilon_A = 1$):

$$F(\varepsilon_A, \varepsilon_B) = \frac{\varepsilon_B(\varepsilon_A-\varepsilon_C)^2}{2(\varepsilon_A+\varepsilon_B)(\varepsilon_A+\varepsilon_C)(\varepsilon_B+\varepsilon_C)}. \tag{4.8}$$

Two simple cases are considered as follows.

1. Solid ε_C is a conductor

This particular case, in which ε_C is infinitely large, leads to

$$F = \frac{\varepsilon_B}{2(\varepsilon_A+\varepsilon_B)}. \tag{4.9}$$

When $\varepsilon_B = \varepsilon_A$ $(= 1)$, $F = 1/4$, and thus Eq. 4.8 results in $A = \pi/3$ and $n = 2/3$. This configuration corresponds to a conductor with a right-angled edge existing isolated in space. When ε_B is much larger than ε_A, on the other hand, $F = 1/2$, and thus $A = \pi/4$ and the lower value of n is equal to 1/2. This means that the presence of a solid dielectric may enhance the field singularity more than in the case of an isolated conductor with no solid dielectric.

2. Dielectric ε_B is supported by the same material

As $\varepsilon_B = \varepsilon_C$ in this case, Eq. 4.8 becomes

$$F(\varepsilon_B, \varepsilon_C) = \frac{(\varepsilon_A - \varepsilon_B)^2}{4(\varepsilon_A + \varepsilon_B)^2}. \tag{4.10}$$

Thus,

$$\cos\left(\frac{n\pi}{2}\right) = \frac{\varepsilon_B - \varepsilon_A}{2(\varepsilon_A + \varepsilon_B)}. \tag{4.11}$$

This equation means that when ε_B changes, the right-hand side varies between 0 and 1/2, and, as a result, the solution of n is always between 2/3 and 1. The case of n being 1 corresponds to $\varepsilon_B = \varepsilon_A$ $(= 1)$. It is self-evident that no field singularity occurs in this case. In the other cases, n is always less than 1, thus causing an infinitely high field at the contact point. With increasing ε_B, n becomes lower, thus enhancing the field singularity.

3. General cases

In general cases other than (1) and (2), the right-hand side of Eq. 4.7 always lies between 0 and 1/2, which leads to the value of n being always between 1/2 and 1. This means that, in contrast to the case in which a solid dielectric supports a plane conductor or electrode, the perpendicular support of a solid dielectric in practice always leads to the occurrence of the infinitely high field strength at a contact point. This characteristic includes case (2) explained above in which the support is of a material with the same dielectric constant. This result is very important, in particular, in high-voltage insulation.

4.3 Numerical Analysis of Field Behavior

4.3.1 Computation of n

First, we compare the value of n obtained from Eq. 4.1 or 4.2 with that resulting from numerical field analysis. Figure 4.4 shows the configuration used in the analysis, the practical meaning of which is explained in Section 5.3. It consists of three dielectrics under a uniform electric field that form a triple junction at the contact point or the interface edge. The dielectric constants (relative permittivities) of the three media are 1.0, 2.3 and 6.0, which correspond to a gas (or a vacuum),

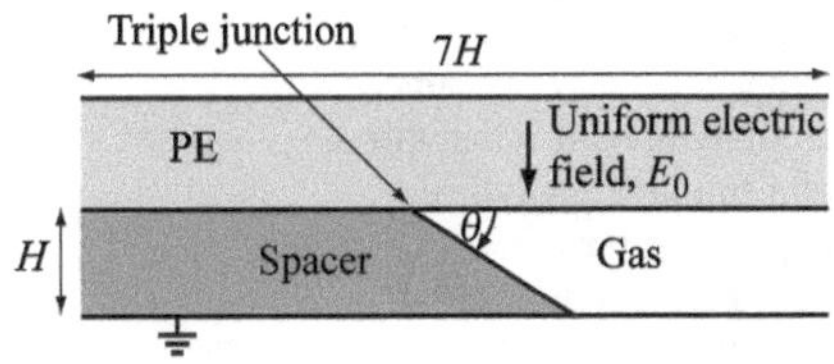

Fig. 4.4 Three-dielectric configuration used for numerically calculating the electric field. *PE*, polyethylene

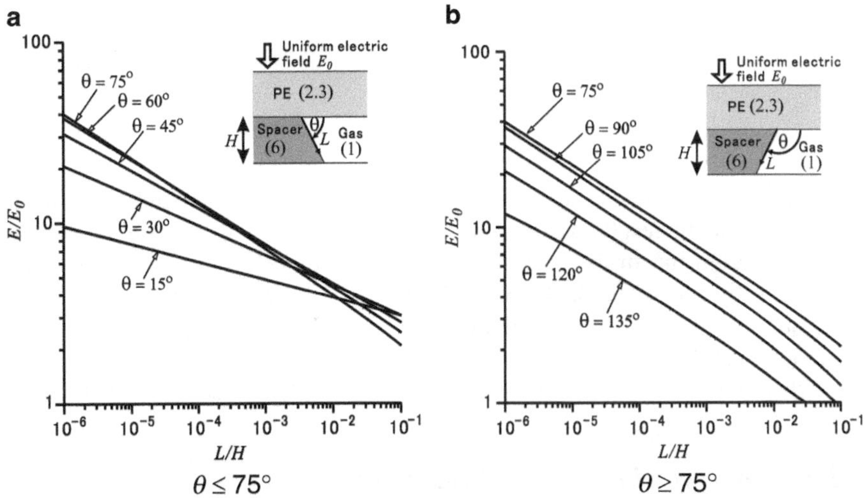

Fig. 4.5 Field strength distribution on the gas side along the spacer surface. The dielectric constant of each medium is given in parentheses [2, 3]

polyethylene (PE) and epoxy resin (indicated as "spacer" in the figure), respectively. The thickness of the epoxy spacer is designated H, which is a unit (normalizing) length in reality. In contrast, the thickness of PE is assumed to be infinitely large as the applied field is uniform.

The calculation was first performed to compute the electric field distribution for various contact angles and dimensions by applying the charge simulation method (CSM) in two-dimensional (2D) conditions. Figures 4.5a and b show the computed (normalized) field strength versus normalized distance L/H (L is the distance along the interface from the contact point) for angles θ below and above 75° (roughly 1.31 radians), respectively. All the curves are straight lines near the contact point on log-log scales, as shown in the figure. This characteristic confirms that the field strength varies as r^{n-1} and its slope gives the value of $(n - 1)$. It should be noted that n is smaller than 1 and $(n - 1)$ is negative. It is also clear that in contrast to a solid dielectric contacting a plane conductor, a right-angled contact with another plane dielectric results in a field singularity.

Figure 4.6 compares the values of $(1 - n)$ thus obtained from the slope of the numerical field calculation (CSM) with those from the analytical expressions.

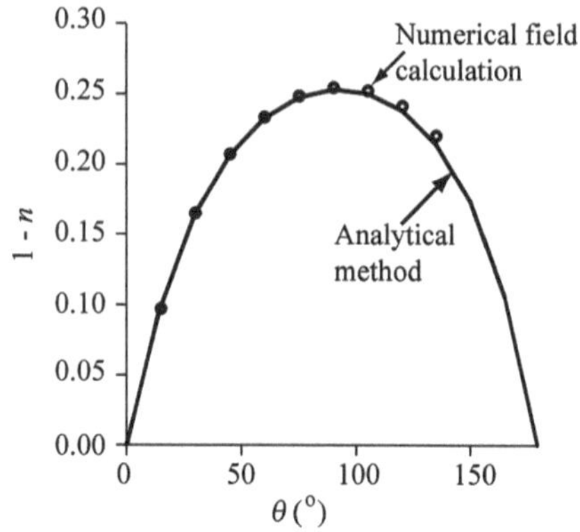

Fig. 4.6 Comparison of exponent $(1 - n)$ obtained from numerical calculations and from the analytical expression for the configuration of Fig. 4.4. ε_A (gas) = 1, ε_B (PE) = 2.3, ε_C (epoxy resin) = 6 [1, 3]. © 2007 IEEE

The latter was computed by applying the Newton-Raphson method to Eq. 4.1. It can be seen that both results agree quite well, thus confirming the validity of the analytical treatment. In the case of $\theta = 45°$ ($\pi/4$ radians), for example, the value of n given by Eq. 4.1 is 0.74686, while the value based on the numerical field calculation is 0.74547. The analytical value becomes slightly lower than the numerically obtained value with increasing contact angle. This is probably due to the difficulty of the numerical field calculation for higher values of angle θ.

4.3.2 *Contact with a Curved Interface*

A variety of configurations exist in practice in which one or two of the three solid dielectrics has a curved contour, thus forming a zero contact angle. Chapters 6 and 7 deal with the field behavior for such configurations as a solid dielectric sphere resting on a plane dielectric, a conductor covered with a solid dielectric layer, a chain of contacting dielectric spheres, and so on. This section explains the field behavior only for a curved interface in the simple configuration of a perpendicular contact under uniform field E_0, as shown in Fig. 4.3. Other results for nonperpendicular contacts are considered in Section 5.3.2.

As can be seen in Fig. 4.6, the absolute value of the exponent $|m (= n - 1)|$ or $(1 - n)$ declines to zero when the contact angle approaches either zero or 180° (π radians). This suggests that a solid dielectric shape rounded to form a zero (or 180°) contact angle may mitigate or diminish the field singularity there. Two variations are possible with such rounding, i.e.,

(a) Inward rounding (zero contact angle), and
(b) Outward rounding (180° contact angle)

and these are shown in Figs. 4.7a and b, respectively. In both configurations, the surface shape of the solid dielectric is rounded to form a quadrant of a circle with radius R near the interface edge (contact point), so as to make a zero (or 180°) contact angle.

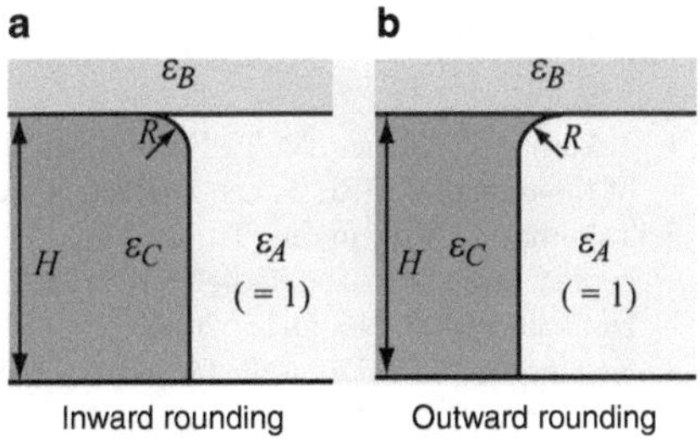

Fig. 4.7 Quadrantal rounding near the point of contact for the configuration given in Fig. 4.3

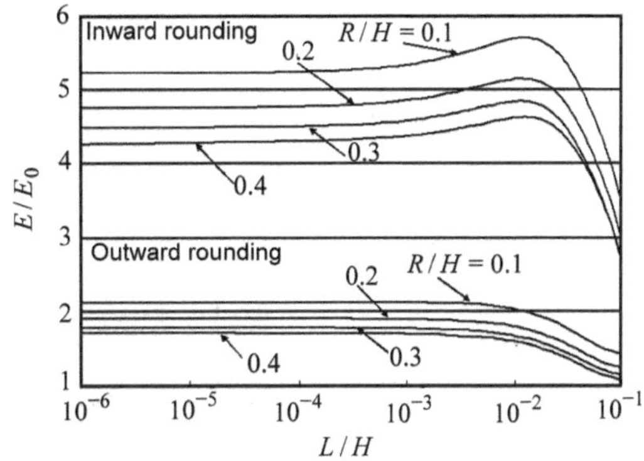

Fig. 4.8 Field behavior on the ε_A side along the interface near the contact point for the configuration given in Fig. 4.7 ($\varepsilon_A = 1$, $\varepsilon_B = 2.3$, $\varepsilon_C = 6$) [3]

Figure 4.8 demonstrates the normalized field strength E/E_0 on the ε_A (gas) side along the interface as a function of the normalized distance L/H (L is the distance along the interface from the contact point) on log-log scales, for inward rounding and outward rounding when these configurations are subjected to uniform field E_0. The dielectric constants used in the computation are as given in the figure legend, corresponding to those in Fig. 4.6. These calculations were also performed by the CSM for various values of rounding radius R/H. From the figure, the following conclusions can be drawn:

(a) Although a straight interface without rounding (corresponding to $R/H = 0$) produces a linearly increasing field strength on log-log scales as the contact point is approached, the rounding mitigates such field concentration. As a result, the field strength becomes almost constant near the point of contact in the range of L/H below about 10^{-3}.
(b) A larger rounding radius R/H makes this constant field lower, in particular, for inward rounding.
(c) Outward rounding results in higher levels of mitigation than inward rounding does, thus lowering the field strength near the contact point. The field strength for inward rounding is two or three times that for outward rounding at the same value of L/H.

References

1. Takuma, T., Kawamoto, T.: Field enhancement at a triple junction in arrangements consisting of three media. IEEE Trans. Dielectr. Electr. Insul. **14**(3), 566–571 (2007)
2. Takuma, T., Kawamoto, T., Goshima, H., Shinkai, H., Fujinami, H.: Field behavior near a contact point of two solid dielectrics. Proc. 13th ACED (Asian Conf. on Electrical Discharge), Hokkaido Univ.:No. O-15, 2006
3. Kawamoto, T., Takuma, T., Goshima, H., Shinkai, H., Fujinami, H.: Triple-junction effect and its electric field relaxation in three dielectrics. IEEJ (Inst. Electr. Eng. Japan), Trans. FM **127** (2), 59–64 (2007) (in Japanese)

Chapter 5
Electric Field in High-Voltage Equipment

Introduction

Chapters 2–4 explain basic field behavior, focusing on the enhancement of the field strength which might take place at the contact point of a solid insulating support. In this chapter, we deal with some practical examples concerning such field behavior near the point of contact in insulation systems. As the content covers several diverse topics, historical overviews are given in each topic area when considered necessary. The main topics covered in this chapter are:

(a) For a finite contact angle, the shape or profile of a solid insulating support (a so-called "spacer") to prevent the field enhancement is discussed. Optimizing calculation of the spacer shape and experimental studies for gas-insulated cables are also briefly mentioned.
(b) For a zero contact angle, the field behavior in various practical configurations is analyzed.
(c) Field behavior and its application to the common contact of three dielectrics are considered.

Additionally we also briefly mention some applications to devices used for high field emission and beam generation.

5.1 Finite Contact Angle: Prevention of Field Singularity near a Contact Point

5.1.1 Field Distribution of a Disc-type Spacer in Coaxial Structures

Discharge inception is governed by electric field distribution in most high-voltage insulation systems. In equipment insulated with SF_6 gas, in particular, the maximum

T. Takuma and B. Techaumnat, *Electric Fields in Composite Dielectrics and their Applications*, Power Systems, DOI 10.1007/978-90-481-9392-9_5,

field strength is decisively important, because discharge usually starts at the point of maximum field strength in SF_6.

It is well known that on the assumption that the radius R_1 of the outer electrode (the so-called "sheath") is kept constant in coaxial electrode configurations, the maximum field strength is lowest when R_1 is about three times (precisely 2.72 times) as the radius of the inner electrode or energized conductor. When this ratio is 3, field E in the gap is given by

$$E = V/(r \ln 3). \tag{5.1}$$

where r is the distance from the central axis and V is the potential difference, which, in practice, is the applied voltage on the conductor. Equation 5.1 means that in a coaxial gas gap, the maximum field strength appears on the conductor surface, and is three times higher than the field strength on the sheath.

Solid insulators, or spacers, are always necessary to support a conductor (often energized at a high voltage) in coaxial structures. For a high-voltage cable insulated with SF_6 gas, the breakdown (or sparkover or flashover) voltage along the surface of a spacer may be much lower than the value in the gas gap itself, and this will be the weak point of the gas insulation system. T. Takuma and T. Watanabe [1] reported the field distribution calculated by the charge simulation method (CSM) for a disc-type spacer in the coaxial cylindrical configuration shown in Fig. 5.1. The spacer has a straight sectional shape, which is symmetrical with respect to both the z-axis and the $z = 0$ plane. An extensive computation was made for the following parameters:

(a) $R_2/R_1 = 1/2$, 1/3, and 1/4
(b) $W/R_1 = 2/15$–$2/3$ (for $R_2/R_1 = 1/3$)
(c) $\alpha = 40°$–$80°$, and
(d) $\varepsilon_s = \varepsilon_B/\varepsilon_A = 2$, 4, and 6

where ε_A and ε_B are the dielectric constants (relative permittivities) of SF_6 gas and the spacer, respectively.

Among these results, Fig. 5.2 shows three typical distributions of normalized field strength on the gas (ε_A) side versus normalized axial distance along the spacer surface for contact angles $\alpha = 50°$, 70°, and 90°. Evidently, the contact angle with

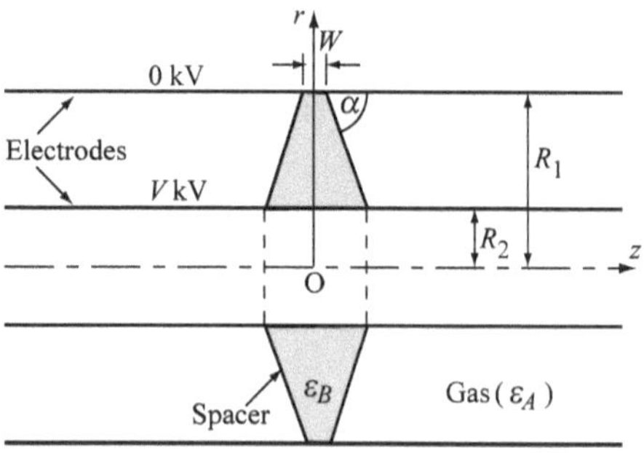

Fig. 5.1 Spacer-electrode configuration used in the calculation

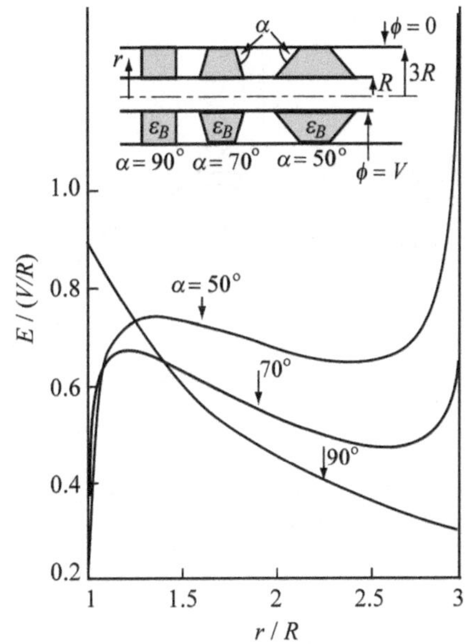

Fig. 5.2 Electric field distribution on the gas side along the spacer surface ($R_2/R_1 = 1/3$, $W/R_1 =$ 2/15–2/3, $\varepsilon_s = \varepsilon_B/\varepsilon_A = 6$, and $R = R_2$ in Fig. 5.1) [1, 7]

the inner electrode (conductor) is (180°−α). In this figure, the radius of the inner electrode (conductor) R_2 is taken as the normalizing size R on both axes.

A spacer with α = 90° does not distort the original field of Eq. 5.1 ($R_2/R_1 = 1/3$) in the gas gap: the field is identical to that without a spacer, i.e., in the case $\varepsilon_B = \varepsilon_A$. In contrast, the use of a spacer with α = 50° or 70° significantly alters the field distribution, resulting in a steep increase near the outer electrode and a steep decrease near the inner electrode. The basic field behavior is explained in Chapter 2 for the contact of a straight dielectric interface with a conductor surface in two-dimensional (2D) cases. This means that the field strength theoretically becomes infinitely high or zero at the contact point, depending on $\varepsilon_B/\varepsilon_A$ and the contact angle. Although Fig. 5.1 shows an axially-symmetrical (AS) configuration, the field behavior may be treated as being 2D in the close vicinity of the contact point.

5.1.2 Optimization of Field Distribution or Spacer Shape

The basic field characteristics explained in Chapter 2 indicate that the original high field strength near the inner electrode in a gas gap may be reduced to zero by using a spacer shape having a contact angle larger than 90° (π/2 radians), as shown in the field for a spacer with α = 50° or 70° in Fig. 5.2. On the other hand, the steep increase in the field near the outer electrode (sheath) can be avoided by adopting a contact angle equal to or larger than 90° there. Thus, it is possible to achieve a field distribution more uniform along the spacer surface (dielectric interface) than that of

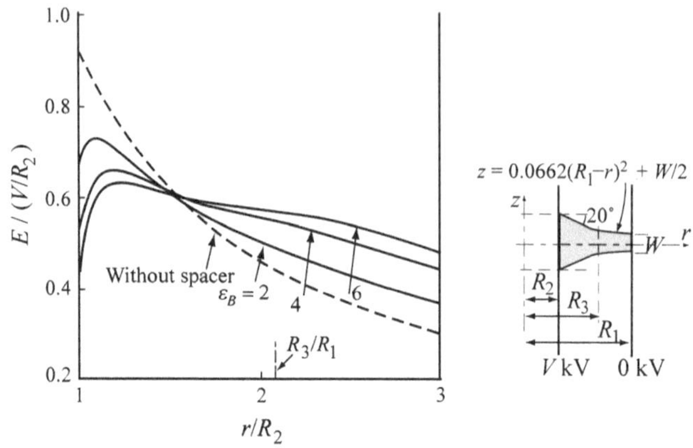

Fig. 5.3 Electric field distribution on the gas side along the spacer surface ($R_2/R_1 = 1/3$, $W/R_1 = 2/15$, $R_3/R_1 = 35/36$) [1]

the original field of Eq. 5.1. T. Takuma and T. Watanabe [1] call spacer shapes based on this idea an optimal profile, and report a detailed study on such field mitigation in coaxial configurations.

One such shape is shown in the right part of Fig. 5.3. Its profile (cross section) is defined by a straight line of $\theta = 20°$ in the range $r/R_1 = 1/3$–$25/36$ and by a quadratic relation $z = 0.0662(R_1 - r)^2 + W/2$ in the range $r/R_1 = 25/36$–1, which smoothly connect with each other at $r/R_1 = 25/36$. For a spacer with dielectric constant $\varepsilon_B = 2$, 4, or 6 ($\varepsilon_A = 1$), the maximum field strength at the inner conductor is reduced due to the presence of the spacer by 20–30% of the original maximum value of Eq. 5.1. They also report an experimental study on the breakdown characteristics for a spacer of polymethyl-methacrylate ($\varepsilon_B = 3.6$) or epoxy resin ($\varepsilon_B = 5.4$), that was performed to verify the effect of field mitigation in air at atmospheric pressure and in compressed SF_6 [1].

The optimizing computation may be extended to realize a constant or uniform field distribution over the entire spacer surface. As far as we know, K. Antolic was the first person to attempt the optimization of a spacer shape (profile or contour) by a numerical field calculation method. He applied the finite element method (FEM) to realize the shape of a cone-type spacer with constant tangential field E_t along the surface [2]. After this work, several papers were published on the optimization of spacer shapes, all of which are based on an iterative procedure of shape transformation and field computation. A short review is given by T. Takuma et al. [3], and more detail is provided in Chapter 19 of T. Takuma and S. Hamada's book [4]. Figure 5.4 is an example of a 3D computation, in which the field (total field strength) is optimized by using the boundary element method (BEM) along a post-type spacer with an elliptical cross section.

It should be noted here that the condition of a constant field on the spacer surface inevitably requires the contact angle to be 90° at both points of contact with the

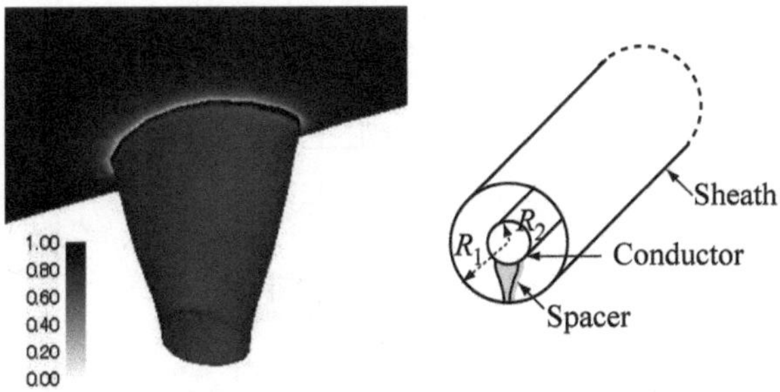

Fig. 5.4 Optimized shape (profile) of a spacer with an elliptical cross-section (the sheath is omitted in the figure on the left). Electric field normalized by the maximal value is expressed by the grey scale on the left figure for $R_2/R_1 = 1/3$ and $\varepsilon_B/\varepsilon_A = 6$ [5]. © 2004 IEEE

electrodes. Furthermore, from the standpoint of the discharge characteristics in SF_6 or any other insulating medium, the following points have not yet been clarified:

(a) Which is the decisive factor, the absolute (total) field E_a or the tangential field E_t?
(b) Which is more desirable near a point of contact, a constant field or zero field strength?

The former problem was studied by H.H. Däumling [6] by comparing the flash-over voltage of a solid insulator between a sphere and a plane, when the shape of the insulator was optimized for E_a and E_t, respectively. However, the experiment was carried out in air at atmospheric pressure and room temperature.

5.2 Zero Contact Angle in Gas-Insulated Equipment

5.2.1 Basic Field Behavior at a Point of Contact

As mentioned in Chapter 3, a zero contact angle means that a solid dielectric and a conductor have a common tangent at the point of contact, and for that, at least one of them must be curved at the contact point. Sections 3.1–3.4 explain the basic properties of contact-point fields for various fundamental configurations, from which the following approximate expressions are presented as contact-point field strength E_C for a zero contact angle,

$$\frac{E_C}{E_{C1}} = f(\varepsilon_s). \tag{5.2}$$

$$f(\varepsilon_s) = \frac{(\varepsilon_s^k + 1)\varepsilon_s}{2} \text{ on the } \varepsilon_A \text{ side, and}$$

$$f(\varepsilon_s) = \frac{\varepsilon_s^k + 1}{2} \text{ on the } \varepsilon_B \text{ side,} \tag{5.3}$$

Fig. 5.5 Two configurations with zero contact angle used in the field calculations [7]. © 1984 IEEE

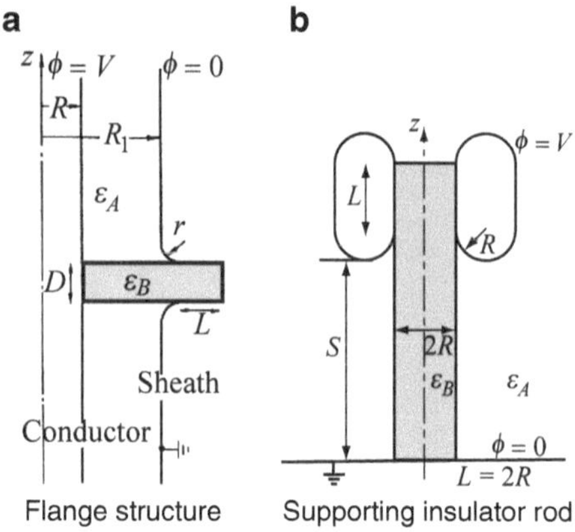

where ε_s is the ratio $\varepsilon_B/\varepsilon_A$ of the dielectric constants between the solid ε_B and the surrounding medium ε_A. E_{C1} is the strength at the corresponding point for $\varepsilon_s = 1$, i.e., the value without a solid dielectric and k is a constant between 0 and 1 in most cases.

There are many possible configurations that include a zero contact angle. This section considers two of them often encountered in practice, as shown in Figs. 5.5a and b. One is a flange structure at the outer electrode (sheath) in coaxial systems, and the other is a supporting insulator rod. Both configurations are axially symmetrical, and the CSM was applied for numerical field calculation.

5.2.2 *Field Behavior in a Flange Structure*

Our interest in the flange structure of Fig. 5.5a lies in the field behavior along a disc spacer intersecting a sheath that is rounded with radius r in cross section. Field calculations were performed for the following cases [7]:

(a) $r/R = 0.1, 0.2, 0.5$, and 1
(b) $\varepsilon_B = 1, 2, 4, 6, 8, 10$, and 20 ($\varepsilon_A = 1$)

The radius of the inner conductor R is the normalizing size, and the sheath radius R_1 is taken as $2R$ or $3R$, while the disc thickness D is equal to R.

Figures 5.6a and b represent the normalized field strength on the surface of the sheath when $R_1/R = 3$, respectively, for $r/R = 0.1$ and 0.5. The abscissa θ is the angle measured from the contact point, which is shown in the upper space of Fig. 5.6a. The field behavior is hardly affected when R_1/R is reduced from 3 to 2, but the influence of r/R is significantly large. As can be seen from Fig. 5.6, the maximum strength decreases to about one third when r/R increases fivefold from 0.1 to 0.5. The maximum does not appear at the contact point. When $\varepsilon_B = 1$, i.e., when only the electrodes are present without a spacer, the position of the maximum

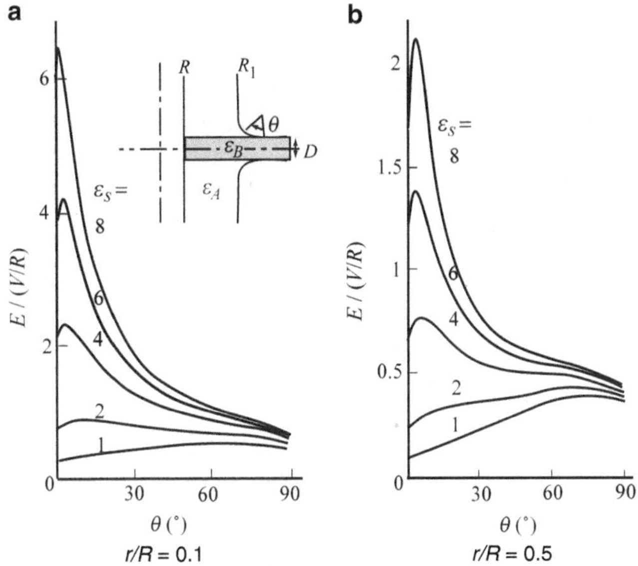

Fig. 5.6 Field distribution along the surface of the outer electrode of Fig. 5.5a ($R_1 = 3R, D = R$) [7]. © 1984 IEEE

Fig. 5.7 Field strength on the ε_B side at the contact point ($\varepsilon_s = \varepsilon_B/\varepsilon_A$)

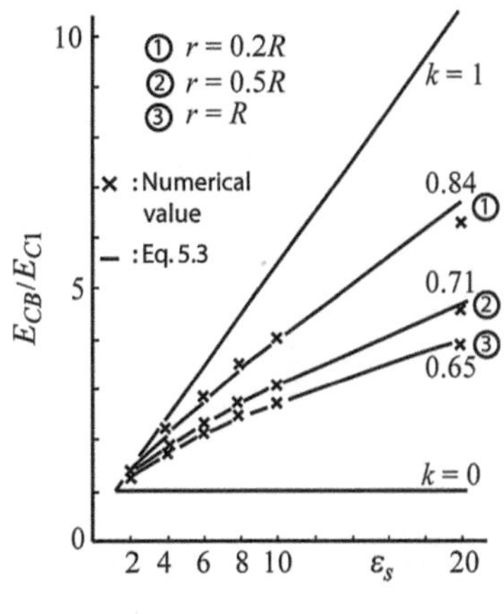

is at $\theta = 70°$ to $80°$, considerably remote from the contact point, but approaches the point with increasing ε_B.

The maximum field strength E_m is about 10–20% higher than the strength E_C at the point of contact, except when E_m appears far away from the contact point. Figure 5.7 shows the field strength E_{CB} on the ε_B (solid dielectric) side at the contact point (normalized by the value E_{C1} for $\varepsilon_s = 1$) together with the line generated using the approximate expression Eq. 5.3. The agreement is fairly good, particularly in the range of ε_B up to 10.

5.2.3 Field Behavior for a Supporting Rod

The second example of a practical configuration is an insulator rod supporting a rounded ring electrode such as a shielding structure used in gas-insulated equipment, as shown in Fig. 5.5b. The parameters used in the calculations are:

(a) $S/R = 2, 4, 8,$ and 16
(b) $\varepsilon_B = 1, 2, 4, 6, 8, 10,$ and 20 ($\varepsilon_A = 1$)

The meaning of these variables is evident from Fig. 5.5b. Radius R of the rod is the same as that of the electrode edge viewed in cross section and L is set at $2R$.

Figure 5.8 presents the (normalized) field distribution on the electrode surface for $S/R = 4$ where the abscissa θ is the angle measured from the contact point. For $\varepsilon_B = 1$, i.e., when the rod does not exist, the maximum field strength appears at θ equal to about 130°. With increasing ε_B, the field strength sharply increases near the contact point, but remains unaffected at larger values of θ above 120°. The local maximum near the contact point exceeds the maximal value appearing at 130° when ε_B reaches 6. This means that for smaller values of ε_B, the ring electrode screens the field enhancement caused by the contact.

The agreement between the contact point field strength obtained from numerical calculation and that using Eq. 5.3 was satisfactory also in the case of Fig. 5.5b [7]. The value of the constant k was found to lie between 0 and 1 in all the cases calculated.

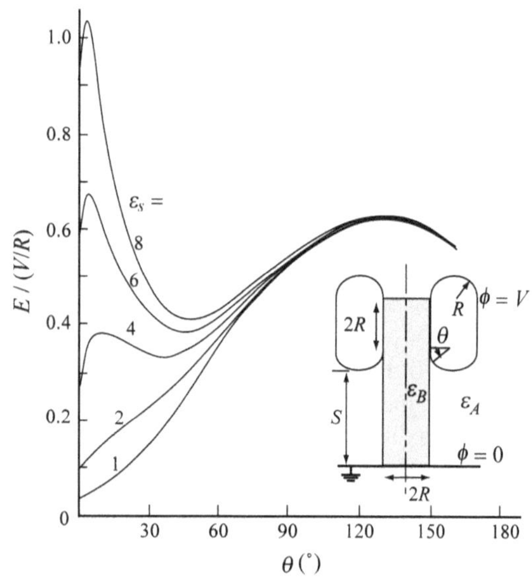

Fig. 5.8 Field distribution along the surface of the ring electrode of Fig. 5.5b ($S = 4R$) [7]. © 1984 IEEE

5.2.4 Other Studies

K. Itaka and T. Hara reported an experimental study on the effect of field enhancement for a post-type spacer in a gas-insulated cable model, as shown in Fig. 5.9 [8]. The configuration was a coaxial one 38/100 mm in diameter. An epoxy-resin spacer having a dielectric constant of about 4 was molded at a right angle on the inner electrode, but fitted in the flange structure of the outer electrode (sheath). The contact edge of the outer electrode was rounded with radius *r* in cross section, where *r* was 1, 3, or 5 mm.

The authors performed flashover tests by applying negative impulse and ac voltages in SF_6 gas at 1–3.5 kg/cm^2 (about 0.1–0.35 MPa) for one case without an insert (shielding) electrode and three cases with an insert, as illustrated in Fig. 5.9. The diameter *d* of the insert was 13.2 or 17.6 mm in cross section, while its height *h* was 7.3 or 11.0 mm. The experimental results showed that in the case without an insert, particularly, when *r* was as small as 1 mm, the flashover voltage was significantly low at the higher pressure of 3.5 kg/cm^2. This low voltage did not occur at the original point of high field (in the gas gap) on the inner electrode, but was initiated due to the field enhancement at the point of contact (point Q in Fig. 5.9) with the outer sheath. An insert electrode mitigates the enhancement there, thus resulting in improved flashover characteristics.

K. Itaka and T. Hara also reported the field distribution calculated by the finite element method (FEM). Figure 5.10 shows the relative field strength on the gas side along the surface of a spacer with an insert for the configuration shown in Fig. 5.9. The maximum field strength appears at the skirt part near the inner electrode, while the value at the contact point near the outer sheath, which is plotted with a dashed line in the figure, is only about one third of the maximum. However, there is a strong possibility that the finite division length in the FEM has resulted in a much lower value than the accurate one in the small gap near the contact point.

K. Itaka et al. also analyzed a similar problem for a cone-type spacer in a coaxial full-size gas-insulated cable [9]. A cone-type spacer of epoxy resin having a dielectric constant of about 6 was molded on a spacer sleeve on the inner electrode (conductor), and was fitted in a flange structure in the outer sheath, as shown in

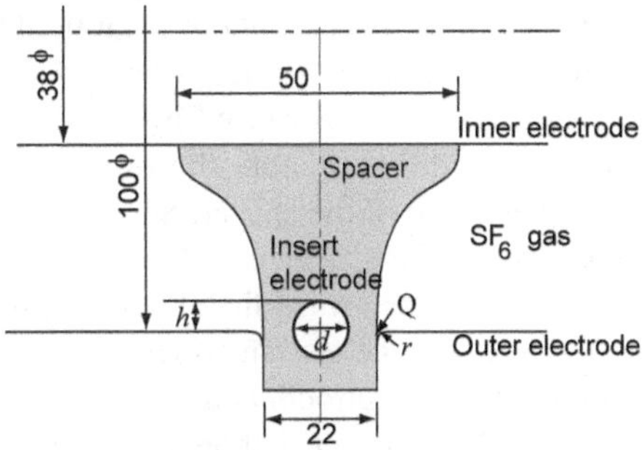

Fig. 5.9 Test configuration of a disc-type spacer between the coaxial electrodes of a cable model. Modified from Fig. 3 in [8]

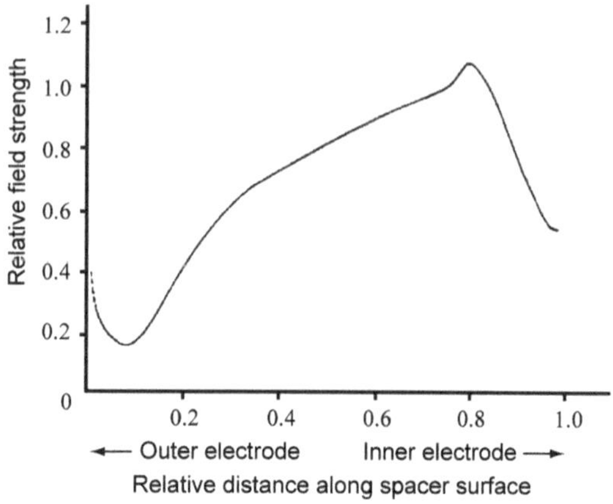

Fig. 5.10 Field distribution along the surface of the disc-type spacer shown in Fig. 5.9 (dimensions of the insert: $d = 13.2$ mm, $h = 7.3$ mm) [8]

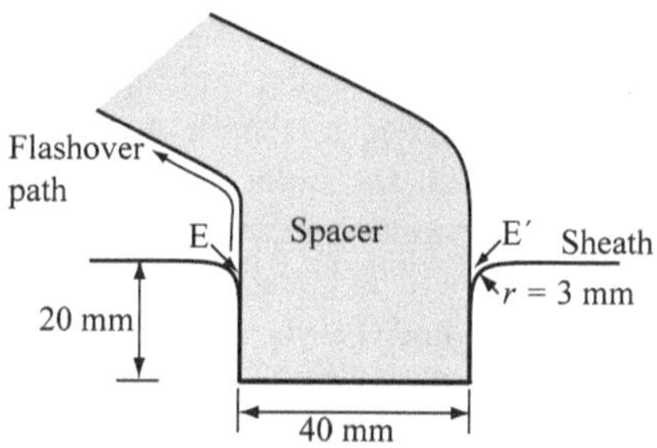

Fig. 5.11 Contact of a cone-type spacer with a sheath [9]. © 1983 IEEE

Fig. 5.11. Because the use of insert shielding electrodes entails such practical disadvantages as manufacturing difficulty and higher cost, they examined various shapes for the spacer and flange, in an effort to improve the electrical performance without using an insert. In the experimental setup shown in Fig. 5.11, no insert electrode was used.

A negative impulse voltage test in SF_6 gas at 3–4.5 kg/cm^2 (about 0.3–0.45 MPa) showed that the flashover voltage was 15–20% lower than the expected value estimated for a spacer without local field enhancement at the contact point. It was found that flashover started from point E near the contact point in Fig. 5.11, suggesting that it was caused by the local field enhancement there. In conclusion, they proposed the structure shown in Fig. 5.12, in which the flange has a small gas gap around part of the spacer. Separation g was 2.5 mm and w was 25 mm in their experiment. The spacer is in contact with the sheath at point Q at a right angle. K. Itaka et al. confirmed the effectiveness of the proposed structure by performing an impulse voltage test, in which flashover did not occur from the sheath side but from the inner electrode, and the flashover voltage was almost as high as the value estimated for a spacer without local field enhancement. The numerical field calculation was also carried out by the FEM for the configuration shown in Fig. 5.11, but

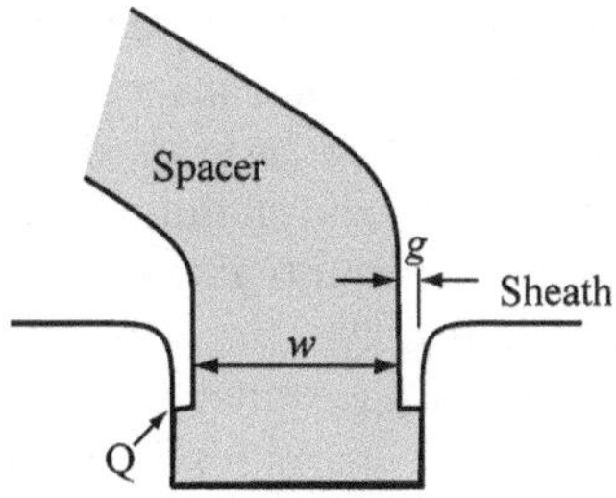

Fig. 5.12 Proposed structure for mitigating field enhancement at a contact point. Modified from Fig. 3 in [9]. © 1983 IEEE

it again seems that the precise estimation was difficult for the steep increase of field strength in the small region near the point of contact.

5.3 Common Contact of Three Dielectrics

5.3.1 Solid Dielectric Supporting Another Solid Dielectric

In contrast to the case of a solid dielectric supporting a conductor explained in Chapter 2, the contact of three dielectrics with cross-sectionally straight interfaces always results in an infinitely high field at the contact point. This field behavior includes cases in which two of the solids are made of the same material. The occurrence of an infinitely high field at contact points is also fully explained in Chapters 2 and 4. In practice, there exist a variety of configurations that involve the contact of three dielectrics, e.g., the contact of two solid dielectrics in a gas or a vacuum.

In summary, the potentials in three media with straight interfaces in the vicinity of the contact point are expressed as a power function as l^n, with electric fields expressed as l^{n-1}, where l is the distance from the contact point. Exponent n is given by solving Eq. 4.1 or 4.2; these two equations are equivalent. Section 4.3 explains the field behavior for a zero contact angle in an originally perpendicular (right-angled) support which is rounded near the point of contact. The effect of such rounding on field mitigation is shown in Fig. 4.8. In this section, we deal with the configurations of a zero contact angle for a non-perpendicular support.

5.3.2 Oblique Solid Surface with a Rounded Edge

The configuration shown in Fig. 5.13, given also in Fig. 4.4, is a simple model of a so-called hybrid gas-insulated transmission line (H-GIL), the intent of which is to take advantage of the insulation characteristics of both SF_6 gas and a solid dielectric [10]. Thus, the original structure is a coaxial one consisting of a central conductor covered with polyethylene (PE) that is supported by an epoxy-resin spacer in SF_6 gas. Field calculations were performed for the simplified model shown in Fig. 5.13, in which all the surfaces (dielectric interfaces) are straight in two-dimensional (2D) conditions, and uniform electric field E_0 was applied. The dielectric constants (relative permittivities) of the three media were ε_A (gas) $= 1$, ε_B (PE) $= 2.3$, and ε_C (epoxy resin) $= 6$.

One of the difficulties associated with the H-GIL is that the use of an epoxy spacer inevitably results in an infinitely high field at a contact point (the triple junction in Fig. 5.13) if the contact angle is finite. As explained in Section 4.3.2, rounding the solid dielectric shape near the contact point to achieve a zero (or 180°) contact angle may relax or mitigate the singularity there. Similarly, as in the case of a perpendicular support, two variations are possible with such rounding for the oblique case:

(a) Inward rounding (a zero contact angle) as shown in Fig. 5.14a, and
(b) Outward rounding (180° contact angle) as shown in Fig. 5.14b

In both cases, the surface shape of the solid dielectric is rounded to form part of a circle with radius R_{SP} near the edge, so as to make a zero or 180° contact angle.

Figures 5.15a shows the normalized field strength E/E_0 on the ε_A (gas) side along the interface (spacer surface) versus the normalized distance L/H (L is the distance along the interface from the contact point) on log-log scales for $\theta = 75°$ for inward rounding. The corresponding results for outward rounding are given in Fig. 5.16a. The results for $\theta = 90°$ (originally perpendicular contact) are also plotted for

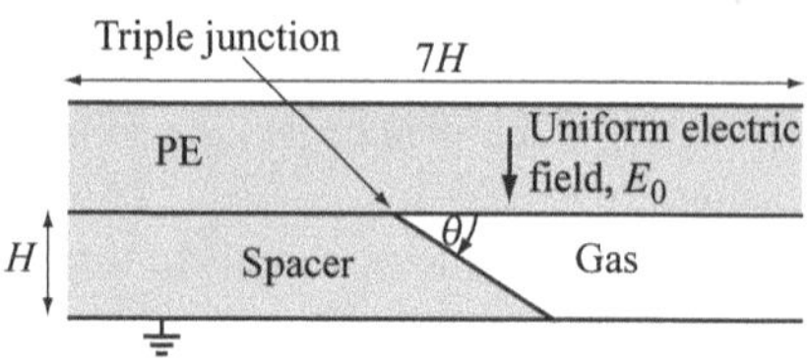

Fig. 5.13 Model configuration of the so-called hybrid gas-insulated transmission line (H-GIL) used in the numerical field calculations. *PE*, polyethylene [10]

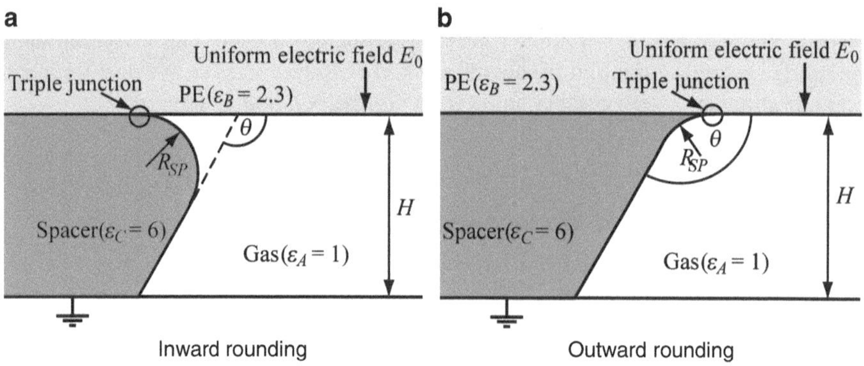

Fig. 5.14 Two configurations of a straight oblique solid surface with a rounded edge at the contact point

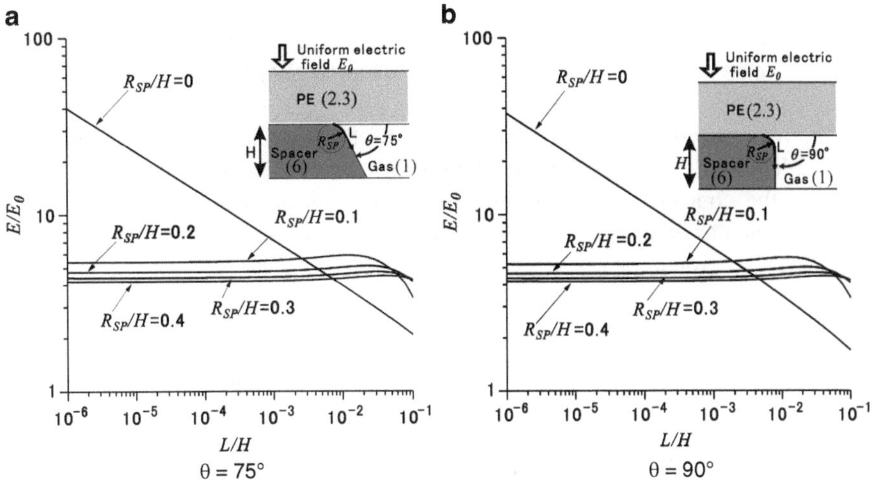

Fig. 5.15 Electric field distribution along the spacer surface for inward rounding as shown in Fig. 5.14a. The dielectric constant of each medium is given in parentheses [11]

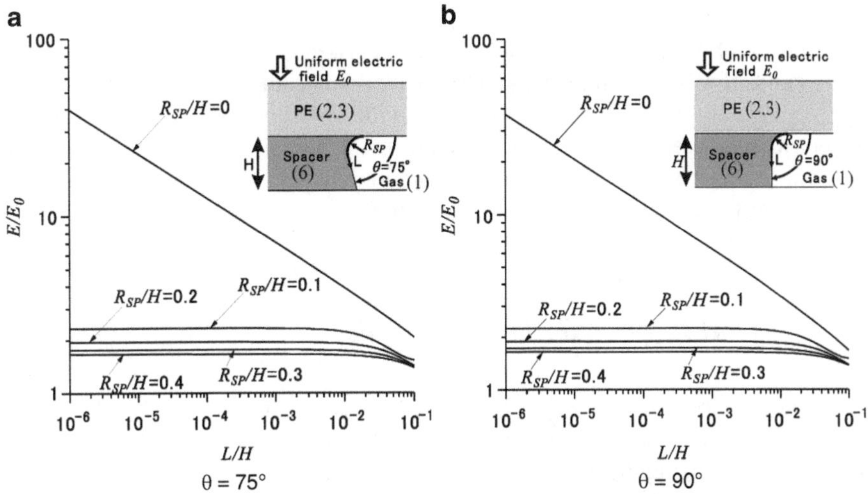

Fig. 5.16 Electric field distribution along the spacer surface for outward rounding as shown in Fig. 5.14b. The dielectric constant of each medium is given in parentheses [11]

comparison in part (b) of both figures. These calculations were performed by the charge simulation method (CSM) for various values of normalized rounding radius R_{SP}/H.

The followings can be concluded;

1. In a similar way to the situation explained in Section 4.3.2, a linear interface without rounding (i.e., $R_{SP}/H = 0$) causes a linearly increasing field strength on

log-log scales as the contact point is approached, but the rounding mitigates this field singularity. As a result of the rounding, an almost constant field strength is realized near the contact point for oblique interfaces, just as for the perpendicular case.
2. The normalized field distribution E/E_0 along the interface for an oblique interface hardly differs from the distribution for a perpendicular interface.
3. Near the contact point, it is the case for oblique interfaces just as it is for perpendicular interfaces that outward rounding results in a lower field strength than inward rounding does.

5.4 Application to High-Field-Emission Devices

The practical importance of analyzing field enhancement is twofold:

(a) Prevention of negative effects leading to possible discharge or breakdown inception, and
(b) Application to devices used for field emission or electron beam generation in a vacuum.

In the latter category, for example, contact between a dielectric interface and a plane conductor is used in diamond surface-emission cathodes in vacuum technology applications [12]. The basic field behavior associated with straight (linear) dielectric interfaces was explained in detail in Chapters 2 and 4. Here, we briefly describe the behavior for two simple configurations and their relationship to the analysis for three dielectrics given in Chapter 4.

5.4.1 *Metal Edge on a Plane Electrode*

The configuration shown in Fig. 5.17 was dealt with, e.g., by L. Schächter [13]. The purpose of the field calculation in this case is either to generate an electron beam in a vacuum by utilizing the field enhancement at the edge or to prevent possible radio frequency (RF) breakdown or arc discharge occurring at the same location.

By comparing Fig. 5.17 with Fig. 4.1, the fundamental figure for three dielectrics meeting at a common point, the corresponding parameters are determined as $\varepsilon_A = 1$,

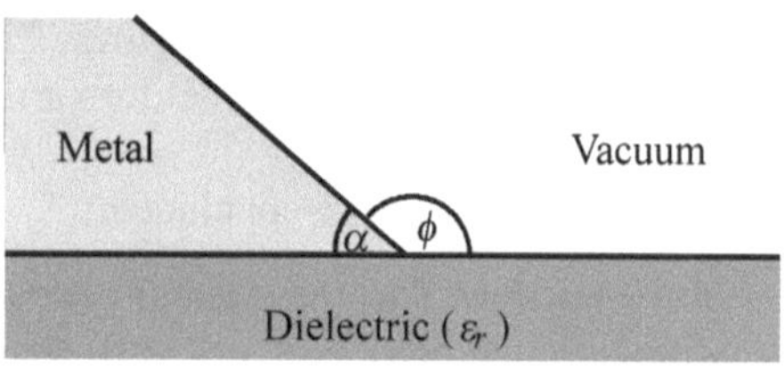

Fig. 5.17 Configuration consisting of a plane dielectric and a metal edge in a vacuum

ε_B = infinity, $\varepsilon_C = \varepsilon_r$, $\theta_A = \phi$, and $\theta_B = \alpha + \phi$, where $\alpha + \phi = \pi$ in Eq. 4.1 or 4.2. The resulting expression is

$$\varepsilon_r \sin A \sin(A - B)\cos(B - C) - \cos A \sin(A - B)\sin(B - C) = 0. \tag{5.4}$$

In this equation, $A = n\theta_A = n\phi$, $B = n\theta_B = n(\phi + \alpha)$, and $C = 2\pi n$. By rearranging Eq. 5.4, the following expression can be derived for determining exponent n:

$$\varepsilon_r \tan n\phi = -\tan n\{2\pi - (\alpha + \phi)\} \tag{5.5}$$

or, because $\alpha + \phi = \pi$,

$$\varepsilon_r \tan n(\pi - \alpha) = -\tan n\pi. \tag{5.6}$$

This expression is the same as the one given in the paper by L. Schächter [13], where v is used instead of n.

5.4.2 General Cases with Two Dielectrics and a Conductor

The configuration shown in Fig. 5.18 was dealt with by M.S. Chung et al. [14]. As for the arrangement given in Fig. 5.17, the purpose of the calculation is the application to generating an electron beam or the prevention of breakdown in a vacuum. By comparing Fig. 5.18 with Fig. 4.1, the corresponding parameters are determined as $\varepsilon_A = 1$, $\varepsilon_B = \varepsilon_r$, ε_C = infinity, $\theta_A = \theta$, and $\theta_B = 2\pi - \alpha$ (or $\theta + \beta$) in Eq. 4.1 or 4.2. The resulting expression is:

$$\varepsilon_r \sin A \cos(A - B)\sin(B - C) - \cos A \sin(A - B)\sin(B - C) = 0, \tag{5.7}$$

where $A = n\theta_A = n\theta$, $B = n\theta_B = n(\theta + \beta)$, and $C = 2\pi n$. Rearranging Eq. 5.7 leads to

$$\varepsilon_r \tan n\theta = -\tan n\beta = \tan(\pi - n\beta). \tag{5.8}$$

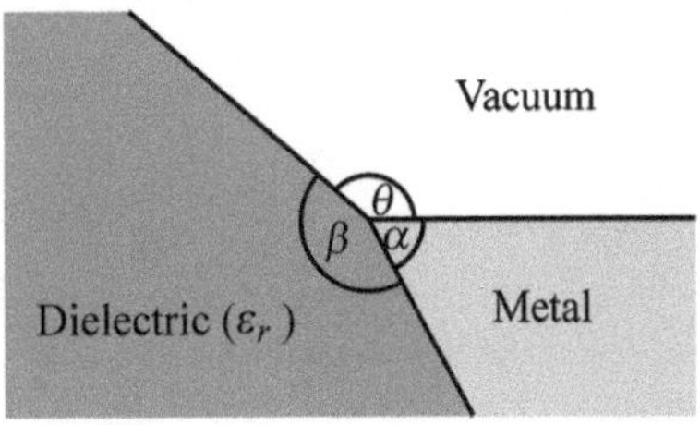

Fig. 5.18 Configuration consisting of a nonplanar dielectric interface and a metal in a vacuum

This expression is the same with the one given in the paper by M.S. Chung et al. [14], where v is used instead of n.

References

1. Takuma, T., Watanabe, T.: Optimal profiles of disc-type spacers for gas insulation. Proc. IEE (GB) **122**(2), 183–188 (1975)
2. Antolic, K.: Die Ermittlung von Randgeometrien mit Hilfe der Differenzenrechnung. Proc. 1st ISH (Int. Symp. High. Volt. Eng.):No.1 (1975) (in German)
3. Takuma, T., Hamada, S., Techaumnat, B.: Recent development in calculation of electric and magnetic fields related to high voltage engineering. Proc. 12th ISH(1), 1–11 (2001)
4. Takuma, T., Hamada, S.: Fundamentals and applications of numerical calculation methods of electric fields, Chap. 19. Tokyo Denki University Press (2006) (in Japanese)
5. Techaumnat, B., Takuma, T., Hamada, S., Kawamoto, T.: Optimization of a post-type spacer in a gas insulated system under three-dimensional conditions. IEEE Trans. Dielectr. Electr. Insul. **11**(4), 561–567 (2004)
6. Däumling, H.H.: Optimization of insulators. Proc. 5th ISH:No.31.95 (1987)
7. Takuma, T., Kawamoto, T.: Field intensification near various points of contact with a zero contact angle between a solid dielectric and an electrode. IEEE Trans. Power Appar. Syst. **103**(9), 2486–2494 (1984), and Takuma, T.: Field behavior at a triple junction in composite dielectric arrangements. IEEE Trans. Electr. Insul. **26**(3), 500–509 (1991)
8. Itaka, K., Hara, T.: Influence of local field-concentration on surface flashover characteristics of spacers in SF_6 gas. Conf. Rec. 1980 IEEE Int. Symp. on Electrical Insulation, IEEE Pub. 80CH 1496–9 ET, 56–60 (1980)
9. Itaka, K., Hara, T., Misaki, T., Tsuboi, H.: Improved structure avoiding local field intensification on spacers in SF_6 gas. IEEE Trans. Power Appar. Syst. **102**(1), 250–255 (1983)
10. Yashima, M., Kawamoto, T., Fujinami, H., Takuma, T., Hata, H., Yamaguchi, H.: Basic properties of hybrid gas-insulated transmission lines (H-GIL). Electr. Eng. in Japan **119**(3), 27–39 (1996), and Yashima, M., Takuma, T., Kawamoto, T.: Basic study on hybrid-gas-insulated transmission line (H-GIL). Proc. 9th ISH:No.7880 (1995)
11. Kawamoto, T., Takuma, T., Goshima, H., Shinkai, H., Fujinami, H.: Triple-junction effect and its electric field relaxation in three dielectrics. IEEJ (Inst. Electr. Eng. Japan) Trans. FM **127** (2), 59–64 (2007) (in Japanese)
12. Geis, M.W., Efremow, N.N. Jr., Krohn, K.E., Twichell, J.C., Lyszczarz, T.M., Kalish, R., Greer, J.A., Tabat, M.D.: Theory and experimental results of a new diamond surface-emission cathode. The Lincoln Lab. J. **10**(1), 3–18 (1997)
13. Schächter, L.: Analytic expression for triple-point electron emission from an ideal edge. Appl. Phys. Lett. **72**(4), 421–423 (1998)
14. Chung, M.S., Yoon, B.-G., Cutler, P.H., Miskovsky, N.M.: Theoretical analysis of the enhanced electric field at the triple junction. J. Vac. Sci. Technol. **B22**(3), 1240–1243 (2004)

Chapter 6
Electric Field and Force in Electrorheological Fluid: A System of Multiple Particles

Introduction

Electrorheological (ER) fluid is a liquid medium that exhibits a significant change in the rheological property when subjected to an electric field [1]. W.M. Winslow discovered in 1949 that some liquids can be immediately solidified when an electric field is applied to them [2, 3]. This phenomenon is called the ER effect or the Winslow effect. Because the viscosity of ER fluids significantly increases in a short time of the order of milliseconds (under a field of a few kV/mm), such fluids may be used in various devices such as dampers, clutches, and actuators [4–8]. The change in fluid viscosity is due to micron-sized particles forming chains in the direction of the applied electric field, which results from the attractive force induced by the electric field.

In the early stages of its research and development, ER fluid was in the form of dielectric particles suspended in a host insulating liquid. At present, ER fluid also exists in the homogeneous form of liquid crystal molecules, in which ER phenomena are caused by the same principles [9]. This chapter discusses the electric field and force in various systems of multiple particles. The particles are treated as spherical, as in the conventional analysis of systems of particles. The basics described here relate not only to ER fluids but also to other applications related to the so-called pearl-chain-forming force with multiple particles.

6.1 Equivalent Dipole Expression

6.1.1 Dielectric Sphere Under a Uniform Field

The electric force exerted on an uncharged dielectric particle is called the dielectrophoretic (DEP) force [10]. The basics of the DEP force may be explained simply using the concept of an equivalent dipole, as shown in Fig. 6.1, in which an uncharged dielectric particle is subjected to electric field $\mathbf{E}_0$. If the dielectric

T. Takuma and B. Techaumnat, *Electric Fields in Composite Dielectrics and their Applications*, Power Systems, DOI 10.1007/978-90-481-9392-9_6,

Fig. 6.1 A dielectric sphere and the equivalent dipole induced by an electric field

constant (relative permittivity) ε_B of the particle is not equal to ε_A of the surrounding medium, polarization will result in the surface (polarization) charge on the particle, which can be represented by equivalent dipole **p** at the particle center. The moment of the dipole is related to $\mathbf{E}_0$ by

$$p = \frac{\varepsilon_B - \varepsilon_A}{\varepsilon_B + 2\varepsilon_A} \times 4\pi\varepsilon_A\varepsilon_0 R^3 E_0, \tag{6.1}$$

where R is the radius of the spherical particle. The direction of **p** is the same as $\mathbf{E}_0$ if $\varepsilon_B > \varepsilon_A$, and opposite to $\mathbf{E}_0$ if $\varepsilon_B < \varepsilon_A$.

The force induced by $\mathbf{E}_0$ on the particle is the sum of that acting on the surface charge, which can be determined exactly from an integral of the stress over the particle surface. The electrostatic force **F** on a particle can be written as the integral of the Maxwell's stress over the particle surface,

$$\mathbf{F} = \varepsilon_A\varepsilon_0 \int \left(\mathbf{E}E_n - \frac{1}{2}E^2\mathbf{a}_n \right) ds, \tag{6.2}$$

where **E** is the electric field on the surface, E is its magnitude (scalar), E_n is its normal component, and $\mathbf{a}_n$ is the unit normal vector on the surface.

However, with the equivalent dipole, the force may be simply expressed as

$$\mathbf{F}_{dip} = (\mathbf{p} \cdot \nabla)\mathbf{E}_0. \tag{6.3}$$

The subscript "*dip*" is used to indicate that this DEP force is estimated from a dipole model. It is clear that the net force arises from the nonuniformity of the electric field applied to the particle. From Eq. 6.3, we can deduce that $\mathbf{F}_{dip}$ directs the particle toward the region of higher field intensity if $\varepsilon_B > \varepsilon_A$. This case is called positive dielectrophoresis. The opposite case, in which the force moves the particle to the region of lower field intensity, is called negative dielectrophoresis. An important note for Eqs. 6.1 and 6.3 is that by using a single dipole **p** to represent the surface charge, the dipole model neglects the surface charge due to field nonuniformity around the particle center.

6.1.2 Multiple Particles

In a system of multiple particles (as mentioned above, we consider particles to be spherical throughout this chapter) under externally applied electric field $\mathbf{E}_0$, the DEP force still occurs even where $\mathbf{E}_0$ is spatially uniform. This is because each

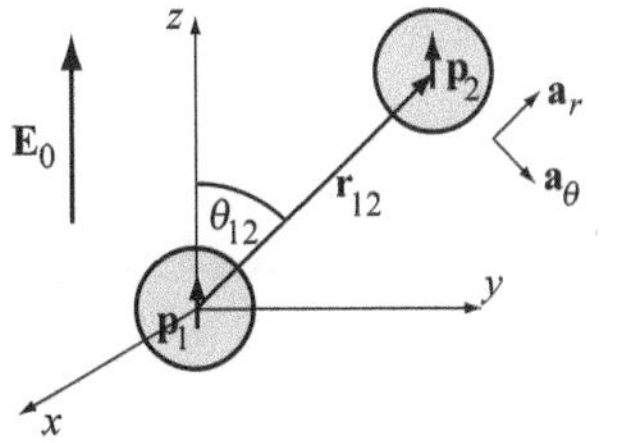

Fig. 6.2 Two dielectric spheres under uniform electric field $\mathbf{E}_0$

particle experiences the nonuniform field caused by the presence of the other particles. From the viewpoint of equivalent dipoles, $\mathbf{F}_{dip}$ is considered to be the force acting between the induced dipole $\mathbf{p}$ of all particles.

We may consider a simple configuration of two spherical particles having the same radius R and the same dielectric constant ε_B in a host medium of dielectric constant ε_A, as in Fig. 6.2. Let particle 1 be at the origin and particle 2 be at spherical coordinates (r_{12}, θ_{12}), as shown in the figure. The applied field $\mathbf{E}_0$ is assumed to be in the z direction. If we use the approximation that the equivalent dipole of each particle depends only on $\mathbf{E}_0$, then

$$\mathbf{p}_1 = \mathbf{p}_2 = p\mathbf{a}_z, \tag{6.4}$$

where p is the dipole moment given by Eq. 6.1.

The DEP force on particle 2 according to this dipole model is determined from p as [11]

$$F_{2,dip} = \frac{3p^2}{4\pi\varepsilon_A\varepsilon_0 r_{12}^4}\left[\left(1 - 3\cos^2\theta_{12}\right)\mathbf{a}_r - \sin 2\theta_{12}\mathbf{a}_\theta\right]. \tag{6.5}$$

On particle 1, the same magnitude of force is exerted but in the opposite direction. One should note that, depending on θ_{12}, the DEP force may attract the particles to each other or repel them from each other. Needless to say, the dipole moment and the force are both zero when $\varepsilon_A = \varepsilon_B$. It is clear that the approximation is appropriate for well-separated particles where $\mathbf{p}_1$ or $\mathbf{p}_2$ of Eq. 6.4 is a good substitution for the induced surface charge.

For closely spaced particles, however, the polarization on one particle significantly modifies the field to which the other particle is subjected. In such cases, the surface charge on each particle cannot be adequately represented by a single dipole, and multipoles must be incorporated [12, 13]. In the following sections of this chapter, the electric field and the force for various configurations of particles are analyzed with an accurate analytical method that includes all multipolar interactions. Although configurations of two dielectric spheres can be analyzed, for example, by using bispherical coordinates [12, 14, 15], the analytical method utilized in this chapter is based on the use of images with multipoles in spherical coordinates. One advantage of this method is that it is applicable to cases of more than two particles. The detail of the calculation method is explained in Section 8.3;

the method used here is more or less similar to the methods applied by M. Washizu and T.B. Jones [13], T.B. Jones [16], and Y. Nakajima and T. Matsuyama [17].

6.2 Particles Lined Up Parallel to an Applied Field

6.2.1 Contact-Point Field

Figure 6.3 shows the two configurations considered in this section. In Fig. 6.3a, a chain of N particles is parallel with uniform external field E_0. The field behavior for Fig. 6.3b is equivalent to that for an infinite number of particles under uniform field V_0/R, where R is the particle radius. All the particles have the same dielectric constant ε_B ranging from 1 to 64 times that of the exterior medium ε_A.

First, we will consider a case in which all the media involved are perfect dielectrics. The maximum electric field usually appears at contact point P_1 between particles in Fig. 6.3a and at P_2 between the particle and an electrode in Fig. 6.3b. Figure 6.4 gives the field at P_1 and P_2 as a function of the ratio $\varepsilon_B/\varepsilon_A$ for E_0 equal to

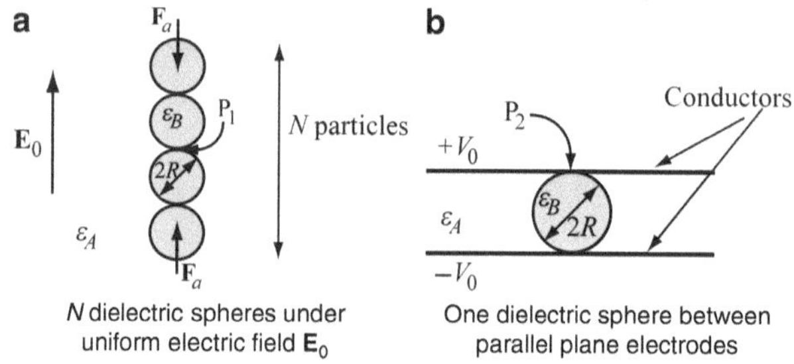

Fig. 6.3 Configurations for analysis

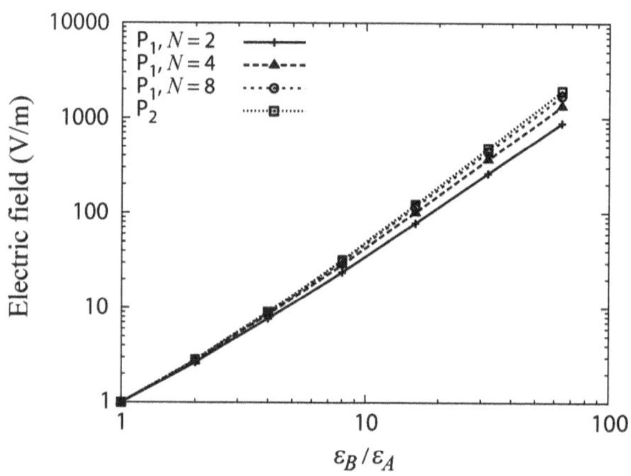

Fig. 6.4 Electric field on the ε_A side at the contact points P_1 and P_2 in Fig. 6.3 [18]. © 2003 IEEE

1 V/m. It is clear that under a uniform field, the contact-point electric field increases when the number of dielectric spheres increases from 2 to 8 in Fig. 6.3a, and to infinity, which corresponds to the configuration shown in Fig. 6.3b. Field enhancement at the contact point with increasing ε_B is clearly demonstrated in Fig. 6.4. However, the slope of each line on the graph only slightly increases with the particle number when using logarithmic scales. This is because the charges on particles located farther from the point P_1 contribute less to the field at P_1.

6.2.2 Approximate Formula for the Contact-Point Field Strength

As explained in Section 3.4.3, the following formula has been proposed for approximating the contact-point electric field E_C on the ε_A side (the host medium):

$$\frac{E_C}{E_{C1}} = \frac{1}{2}\frac{\varepsilon_B}{\varepsilon_A}\left[\left(\frac{\varepsilon_B}{\varepsilon_A}\right)^k + 1\right], \tag{6.6}$$

where E_{C1} is the contact-point field for $\varepsilon_B/\varepsilon_A = 1$, and k is an appropriate real number parameter depending on the geometrical configuration. Equation 6.6 is equivalent to Eq. 3.11, where $\varepsilon_s = \varepsilon_B/\varepsilon_A$. The values of k computed by curve fitting in Fig. 6.4 are given in Table 6.1 for $\varepsilon_B/\varepsilon_A$ values up to 64. Under a uniform electric field, they are smaller than unity, as mentioned in Section 3.4.3 and as predicted by T. Takuma [19]. However, it has been found from analysis of a particle subjected to a nonuniform field that k can be slightly greater than unity in general [18]. Using the values of k in Table 6.1, differences from the analytical results are smaller than 8% in this case, as shown in Table 6.2.

On the particle chain shown in Fig. 6.3a, the DEP force is parallel to the field and constitutes an attractive force among the particles. In other words, the force is upward on particles in the lower half of the chain and downward on those in the upper half. Replacing particles with their equivalent dipoles, we can deduce that the two particles at the ends of the chain are subjected to the strongest force, as the interactions with all the other dipoles yield a force in the same direction.

Table 6.1 Values of k for approximating the contact-point field at P_1 and P_2 using Eq. 6.6 [18]. © 2003 IEEE

Contact point	P_1			P_2
	$N = 2$	$N = 4$	$N = 8$	
k	0.786	0.889	0.946	0.971

Table 6.2 Differences (%) in the electric field between estimations using Eq. 6.6 and analytical results [18]. © 2003 IEEE

$\varepsilon_B/\varepsilon_A$	P_1			P_2
	$N = 2$	$N = 4$	$N = 8$	
4	3.66	4.10	6.28	7.75
16	1.41	1.32	2.79	4.08
64	1.54	0.69	2.27	4.17

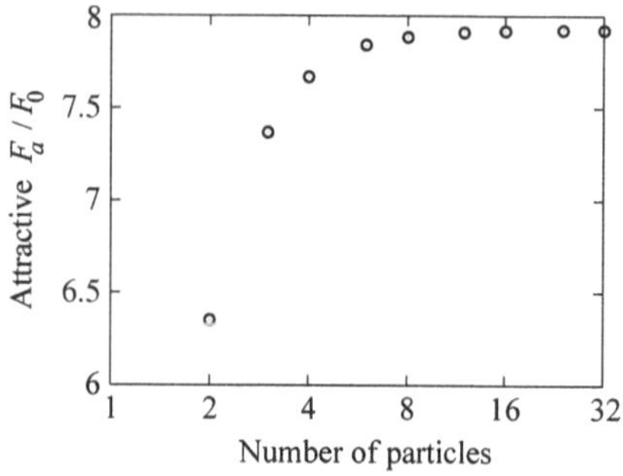

Fig. 6.5 Force on the end particles of the chain shown in Fig. 6.3a ($\varepsilon_B/\varepsilon_A = 4$) [20]. © 2004 AIP

The maximal force F_a (subscript 'a' stands for attractive) normalized by $F_0 = \varepsilon_A\varepsilon_0 {E_0}^2 R^2/2$ is shown in Fig. 6.5 as an example for $\varepsilon_B/\varepsilon_A = 4$ and N up to 32. It is clear from the figure that F_a increases with the number of particles in a chain, but reaches saturation at sufficiently large N, which is about 16–32 for $\varepsilon_B/\varepsilon_A = 4$.

6.3 Particle Chain Tilted to the Field Direction

6.3.1 Chain of Two Particles

This section investigates the force behavior in general cases in which a chain of spherical particles makes an angle with the applied field, and the force behavior on a chain having one end attached to a planar electrode. The latter case is often encountered in practice because image charges with respect to an electrode generally attract a particle chain toward the electrode. In this section, the ratio of dielectric constants is kept constant at $\varepsilon_B/\varepsilon_A = 4$ unless otherwise stated.

Initially, we consider an isolated chain of N particles making tilt angle θ_E with external uniform field $\mathbf{E}_0$ in the vertical direction, as shown in Fig. 6.6. In the case of $N = 2$, the direction and the magnitude of DEP forces $\mathbf{F}_1$ and $\mathbf{F}_2$ are illustrated in Fig. 6.7 for $\theta_E = 0°$, 45°, and 90°. Note that the electric field and force both increase with $\varepsilon_B/\varepsilon_A$. In the figure, $\mathbf{F}_1$ and $\mathbf{F}_2$ are the forces on the lower and upper particles in the chain, respectively. Figure 6.7 demonstrates that both the magnitude and direction of the force depend heavily on the angle θ_E. For $\theta_E = 0°$, $\mathbf{F}_1$ and $\mathbf{F}_2$ are parallel to the chain and are attractive, where $F_1 = F_2 = 6.352F_0$. For $\theta_E = 90°$, $\mathbf{F}_1$ and $\mathbf{F}_2$ are also parallel to the chain but have become repulsive, where $F_1 = F_2 = 0.828F_0$, i.e., much smaller than the force for zero θ_E.

For θ_E between 0° and 90°, the forces are not parallel to the chain, as shown for $\theta_E = 45°$. They are regarded as the sum of two vectors, $\mathbf{F}_a$ and $\mathbf{F}_b$, which are parallel and perpendicular to the chain direction, respectively. Positive or negative F_a values mean that the particle is either attracted to or repelled from the chain. Therefore, it is inferred from Fig. 6.7 that the DEP force changes from attractive to repulsive at a certain transition angle. This angle is about 70° for a two-particle

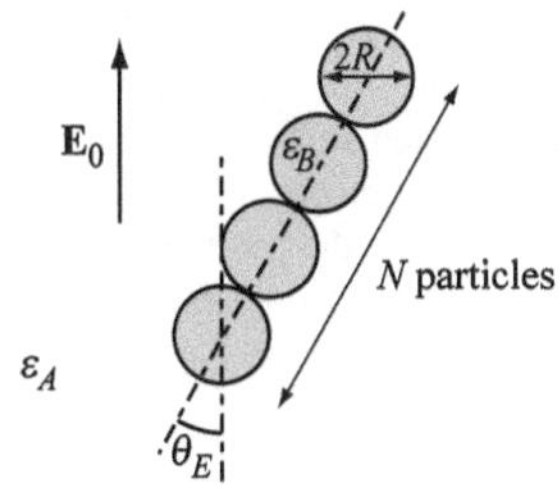

Fig. 6.6 An isolated N-particle chain making angle θ_E with an external uniform field

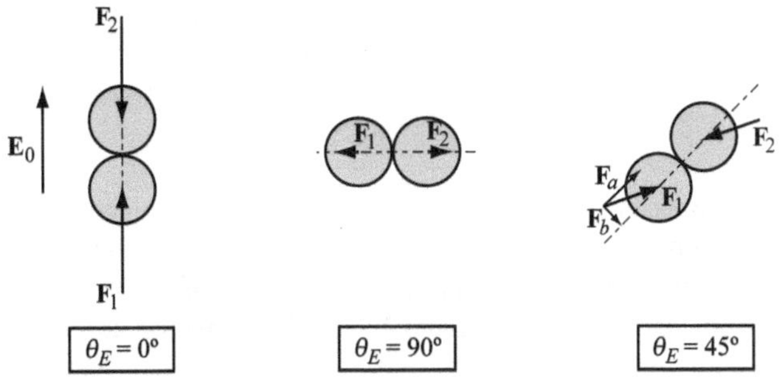

Fig. 6.7 The DEP force on an isolated chain of two particles under uniform vertical field $\mathbf{E}_0$ [20]. © 2004 AIP

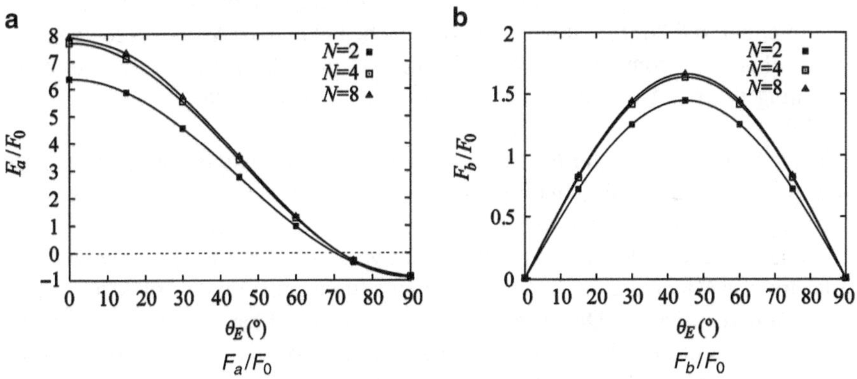

Fig. 6.8 Force magnitude on an isolated chain of N particles making angle θ_E with uniform field $\mathbf{E}_0$ [20]. © 2004 AIP

chain with $\varepsilon_B/\varepsilon_A = 4$. On the other hand, positive F_b is set in the direction of decreasing θ_E, i.e., tending to align the chain with the field.

6.3.2 Isolated Chain of Multiple Particles

For particle number N ranging from 2 to 8 in Fig. 6.6, the behavior of F_a and F_b on the highest and lowest particles in the chain is shown as a function of θ_E in

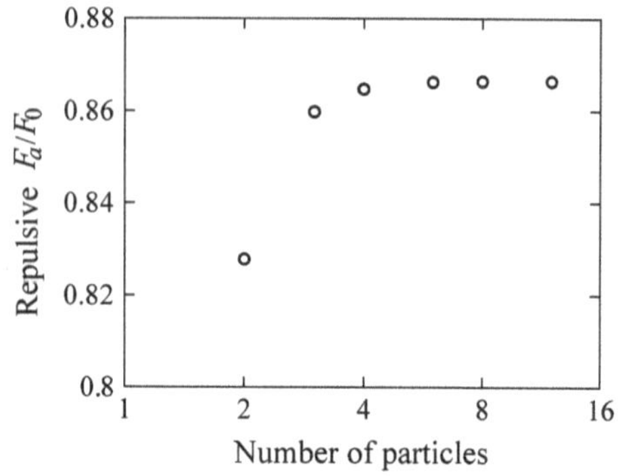

Fig. 6.9 Maximum of the repulsive force at $\theta_E = 90°$ on an isolated particle chain under a uniform field as a function of the particle number N [20]. © 2004 AIP

Figs. 6.8a and b. We can see that the transition of $\mathbf{F}_a$ to a repulsive force takes place at about θ_E between 70° and 73° for all N. This transition angle slightly increases with N, implying that a longer chain is stabilized by the DEP force for a wider range of chain directions. However, the effect of N on the transition angle is negligible at N higher than about 4.

The maximum values of the attractive and repulsive forces occur at $\theta_E = 0°$ and 90°, respectively, as expected. Figure 6.9 shows the increase of the maximal repulsive force with the number N of particles in the chain. A similar figure for the attractive force at $\theta_E = 0°$ has already been presented in Fig. 6.5. The saturation value is equal to $0.866F_0$, which is almost reached at $N = 6$. As the force increases by less than 5% from that for $N = 2$, the two-particle model can be used for estimating the repulsive force for a longer chain, provided the ratio $\varepsilon_B/\varepsilon_A$ is not too high.

In contrast to the case for F_a, Fig. 6.8b shows that the component F_b of the DEP force perpendicular to the chain direction is invariably positive. As a result, an isolated particle chain always tends to become aligned in the field direction by this component of the DEP force. The maximum value of F_b occurs at $\theta_E \approx 45°$, and varies little as N increases from 4 to 8. From the force behavior described above, we may summarize that the DEP force stabilizes an isolated particle chain toward the direction of a uniform field. The force magnitude, maximal on the particles at the chain ends, increases with the number of particles, being more or less invariant for a sufficiently long chain. Such alignment of particles is observed in experiments of free particles or biological cells under an applied field but far away from the electrodes.

6.3.3 *Two-Particle Chain in Contact with a Plane Electrode*

The configuration for analysis is a two-particle chain on a grounded electrode under external uniform field $\mathbf{E}_0$, as shown in Fig. 6.10. The chain is tilted at angle θ_E. The forces $\mathbf{F}_1$ and $\mathbf{F}_2$ on the lower and upper particle, respectively, are illustrated in

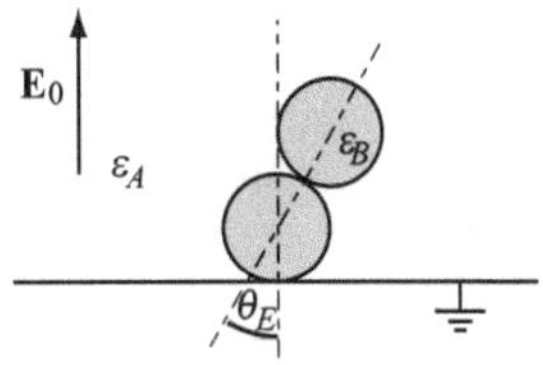

Fig. 6.10 A two-particle chain in contact with a plane electrode under uniform field $\mathbf{E}_0$

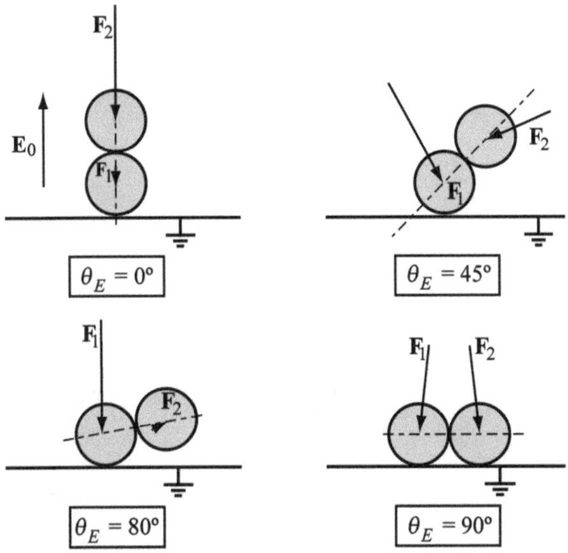

Fig. 6.11 Behavior of the DEP force on a two-particle chain in contact with a grounded electrode under uniform field $\mathbf{E}_0$ [20].

Fig. 6.11. It is clear that the force behavior is significantly different from that in Fig. 6.7, i.e., the configuration without a planar electrode. First, $\mathbf{F}_1$ always attracts the lower particle to the electrode. The magnitude F_1 is quite small, being $0.888F_0$ at $\theta_E = 0°$, but grows considerably with increasing θ_E. The downward component of F_1 (normalized by F_0) is shown as a function of θ_E in Fig. 6.12. Second, the electrode also has a significant influence on force $\mathbf{F}_2$ on the upper particle: we can see an abrupt change in the magnitude and direction of $\mathbf{F}_2$ in Fig. 6.11 when θ_E increases from 80° to 90°. With increasing θ_E, the upper particle moves closer to the electrode; the result is that the weak, upward $\boldsymbol{F}_2$ at $\theta_E = 80°$ becomes a strong, downward force at $\theta_E = 90°$.

A comparison of the DEP force on the upper particle with and without the plane electrode is presented separately for components F_a and F_b in Figs. 6.13a and b. Although the characteristics are similar, F_a is stronger with the electrode than without the electrode. The major difference can be observed for F_b at high θ_E near 90°, i.e., a strong downward attraction develops due to the presence of the

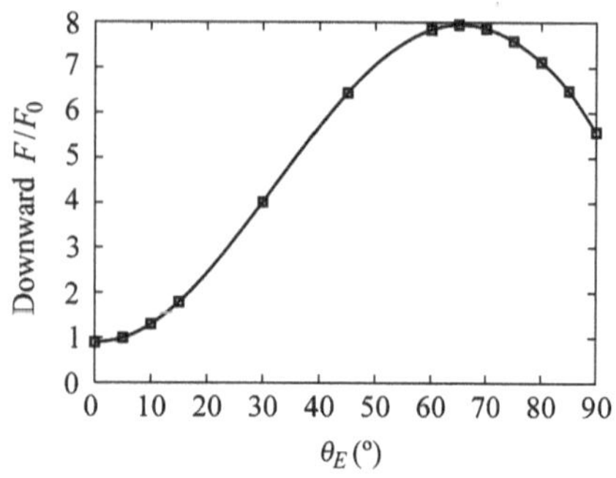

Fig. 6.12 Downward component of the force acting on the lower particle to the electrode in Fig. 6.10 [20]. © 2004 AIP

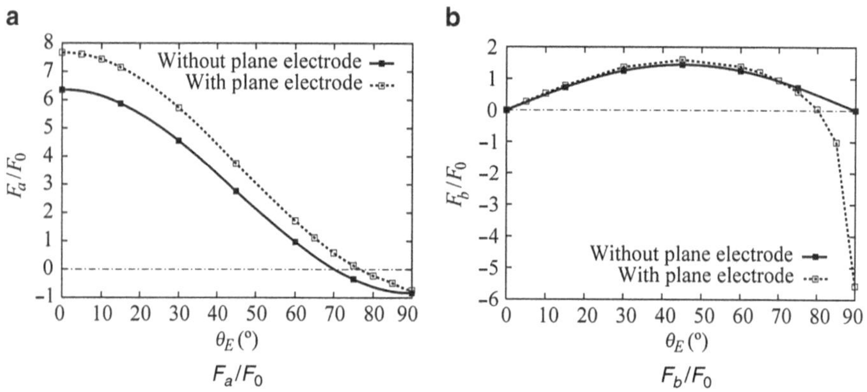

Fig. 6.13 Comparison of the force components F_a and F_b on the upper particle of a two-particle chain for θ_E from 0 to 90° [20]. © 2004 AIP

electrode. In conclusion, the comparison shows that a particle chain is stabilized for a wider range of θ_E and by a stronger force for most values of θ_E when the end of the chain is in contact with an electrode.

6.3.4 Two-Particle Chain Between Parallel Plane Electrodes

We study the effect of the presence of an upper electrode on the force of a two-particle chain in the configuration with two plane electrodes shown in Fig. 6.14. Table 6.3 presents the effect of the upper electrode at various values of electrode separation D on the DEP forces F_1 and F_2 for $\theta_E = 45°$ where $F_0 = \varepsilon_A \varepsilon_0 {E_0}^2 R^2/2$. The table clearly shows both F_1 and F_2 increasing with decreasing D, compared with the values for the configuration shown in Fig. 6.10, which corresponds to the case of infinite D. The effect, however, is comparatively small. For example, even for D/R as small as 4 (i.e., the separation between the top of the upper particle and the upper electrode is only 0.586R), the relative differences ΔF_1 and ΔF_2 are both

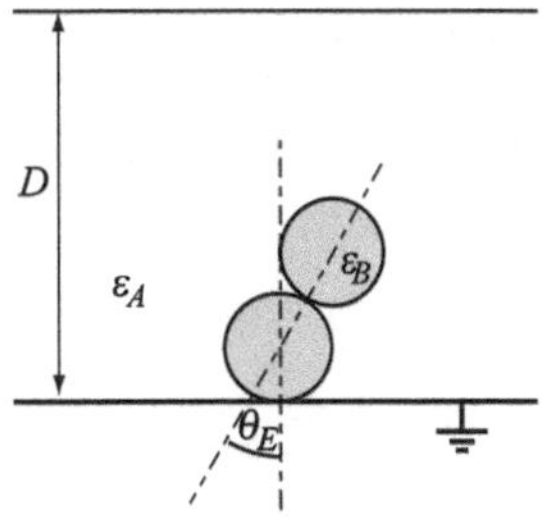

Fig. 6.14 Particle chain attached to the lower electrode between parallel plane electrodes with separation D

Table 6.3 Effect of electrode separation D on the DEP forces for a two-particle chain in the configuration shown in Fig. 6.14 when $\theta_E = 45°$ [20]. © 2004 AIP

D/R	Lower particle		Upper particle	
	F_1/F_0	ΔF_1 (%)	F_2/F_0	ΔF_2 (%)
Infinity	7.463	0.00	4.072	0.00
10	7.481	0.24	4.082	0.25
8	7.498	0.47	4.093	0.51
6	7.552	1.20	4.127	1.35
5	7.638	2.34	4.181	2.67
4	7.860	5.32	4.311	5.87

less than 6%. Therefore, for a particle chain attached to an electrode, the effect of the counter electrode can be neglected in most cases where the electrode is farther away from the chain than the particle radius.

6.4 Two-Particle Chain Between Parallel Plane Electrodes with the Minimum Separation

6.4.1 Scope of the Section

As explained above in Section 6.3.2, the DEP force aligns a particle chain so as to be parallel with an applied electric field, and the influence of an electrode on the force behavior is significant only when a particle is close to the electrode. This section analyzes the restoring force on a two-particle chain between parallel plane electrodes with the separation D set so that the spherical particle chain completely bridges the electrode gap, i.e., $D = 4R$. Figure 6.15 shows the state of equilibrium. In this section also, the ratio of dielectric constants is kept constant at $\varepsilon_B/\varepsilon_A = 4$ unless otherwise stated.

This configuration simplifies real situations in which chains composed of a larger number of particles are formed across an electrode gap by the DEP force. In this situation, the largest effect on the electric field and resultant force arises from

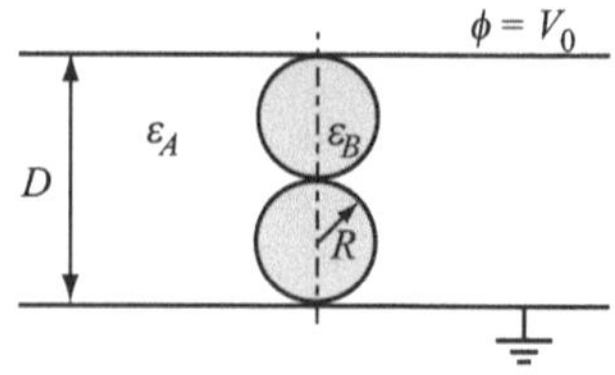

Fig. 6.15 Two-particle chain with both ends in contact with parallel plane electrodes

the interaction between adjacent particles. For the particles at the end of the chain, on the other hand, the electrode also plays a significant role on the field and force behavior. For the configuration shown in Fig. 6.15, the field behavior at the end of the chain mainly approximates that of a particle at the edge of a longer chain, whereas the behavior near the middle of the chain corresponds to that of an intermediate particle in a longer chain.

The focus of this section is on the variation of the DEP force in possible situations where a chain starts tilting to be non-parallel with the applied electric field. As the particle chain is still close to the electrodes, the effect of the electrodes must be fully taken into account in the calculation. Under the equilibrium shown in Fig. 6.15, the net force is zero on both particles because of the symmetry brought by the parallel plane electrodes.

6.4.2 Electric Field Distribution

Let us assume that the upper electrode slides in the horizontal direction to the right under the application of shear or a shearing force, and the particle chain is tilted from the vertical position. In Fig. 6.16 are shown three possible geometrical configurations of the chain for tilt angle θ_E. They are as follows:

(i) The chain is split by distance $d = 2R[(\cos\theta_E)^{-1} - 1]$ so as to remain in contact with the electrodes (Fig. 6.16a).
(ii) The chain rotates so that the particles are still in contact with each other but are separated by the same distance from each electrode (Fig. 6.16b).
(iii) The chain is not separated, but rotates, remaining in contact with the lower electrode (Fig. 6.16c).

Figure 6.17 shows examples of the electric field distribution for $\varepsilon_B/\varepsilon_A = 4$ for each of the above three configurations on the upper particle along a line defined by zenith angle $\theta = 0°$–$180°$ and azimuth angle $\varphi = 270°$, i.e., the thick contour in the figure associated with each graph. Figure 6.17a shows that the maximum field strength in the arrangement of Fig. 6.16a is located at $\theta = 0°$, the contact point between the particle and the upper electrode. This maximal field scarcely varies with θ_E. There is also a local peak of the electric field along the contour at $\theta = 180° - \theta_E$, which is the point closest to the lower particle. Unlike the maximum at $\theta = 0°$, the local peak is reduced significantly on increasing θ_E, or with a wider

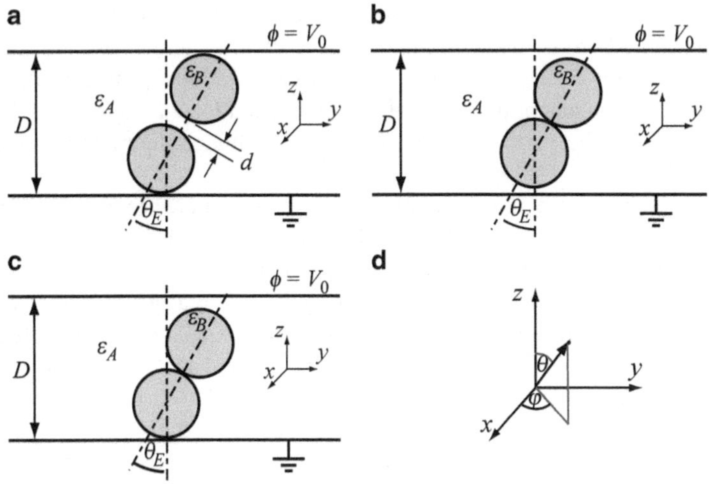

Fig. 6.16 Possible configurations of a tilted two-particle chain under shear

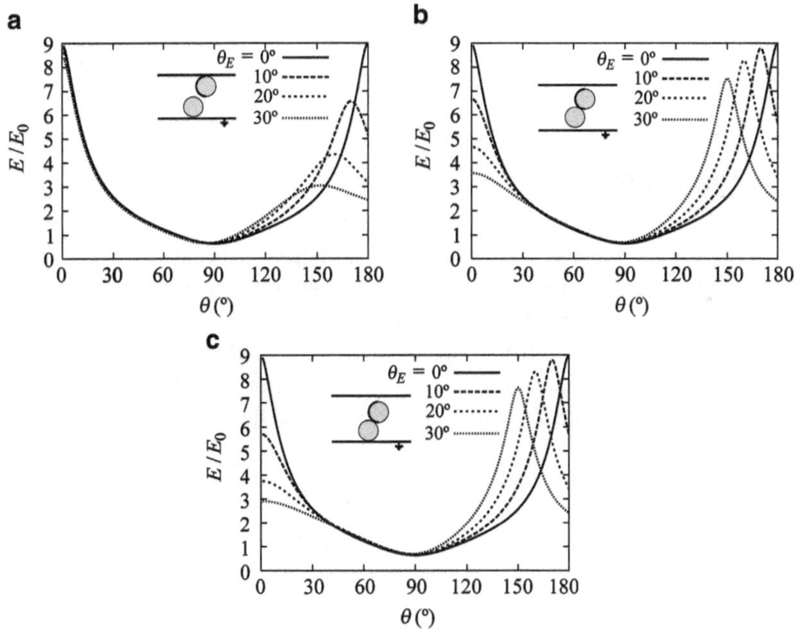

Fig. 6.17 Field distribution on the line $\theta = 0°$ to $180°$ and $\varphi = 270°$ of the upper particle (shown as the *thick contour*) for the three configurations of a tilted two-particle chain under shear [21]

separation between the particles. For higher values of $\varepsilon_B/\varepsilon_A$, the field (not shown here) behaves in a similar way to that indicated in Fig. 6.17a, but it is more intensified in the high-field regions.

The electric field for the particle chain separated by the same distance from each electrode (Fig. 6.16b) is shown in Fig. 6.17b. In contrast to the case for Fig. 6.17a, the maximum field occurs at $\theta = 180° - \theta_E$, whereas a local peak exists at $\theta = 0°$. As θ_E increases, the magnitudes of both the global and local field maxima decrease. For the same θ_E, the maximal field is smaller in this configuration than it is in Fig. 6.16a.

For the upper particle of a chain attached to the lower electrode (Fig. 6.16c), the field distribution is given in Fig. 6.17c and is essentially similar to that in Fig. 6.17b. The main difference is the larger reduction of the electric field at $\theta = 0°$ in Fig. 6.17c with increasing θ_E. It is worth noting that whereas geometrical symmetry of the field distribution exists with respect to the mid-plane in both Figs. 6.16a and b, such symmetry does not apply to Fig. 6.16c. The field distribution on the lower particle is shown in Fig. 6.18 for the configuration in Fig. 6.16c ($\varphi = 90°$ in this case). Clearly, at $\theta = 180°$, i.e., the contact point between the lower particle and the electrode, the field has nearly the same magnitude as that on the upper particle at $\theta = 0°$ in Fig. 6.17a. On the other hand, at $\theta = \theta_E$, the contact point between the particles, the field behaves similarly to that in Fig. 6.17b or c.

It may be summarized from these results that the electric field is more or less invariant with θ_E near the contact point between a particle and a conducting plane, but significantly decreases with increasing θ_E near the contact point between the particles.

6.4.3 DEP Force

Figures 6.19–6.21 present the force on the upper particle of the chains in Figs. 6.16a, b, and c, respectively. A nonzero net force occurs on a chain tilted from the vertical position of equilibrium, and this force can be separated into horizontal component F_y and vertical component F_z, referring to the coordinate axes in Fig. 6.16. The force is normalized by $F_0 = \varepsilon_A \varepsilon_0 E_0{}^2 R^2/2$, where $E_0 = V_0/D$, i.e., the force acting on any $R \times R$ square element on the electrode surfaces in the absence of the chain.

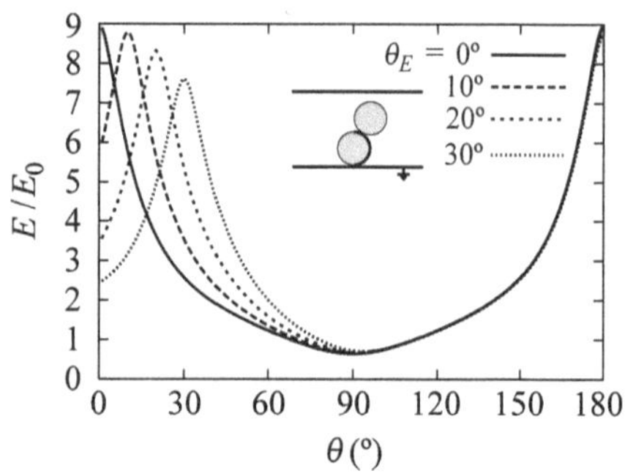

Fig. 6.18 Field distribution on the lower particle at $\varphi = 90°$ (shown as the *thick contour*) for the configuration in Fig. 6.16c [21]

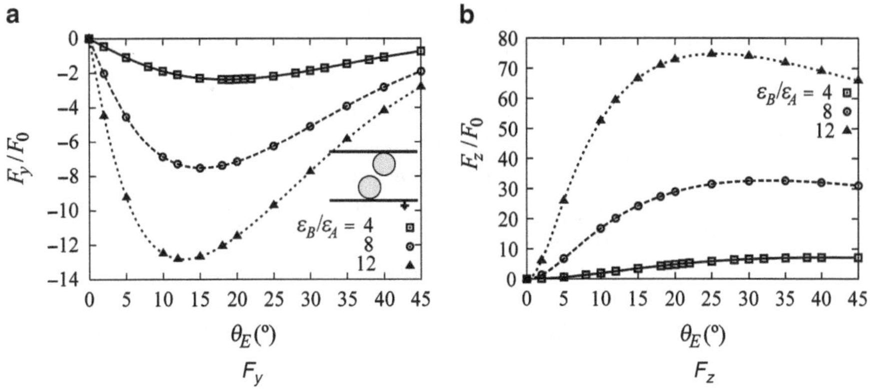

Fig. 6.19 Force components on the upper particle in Fig. 6.16a [21]

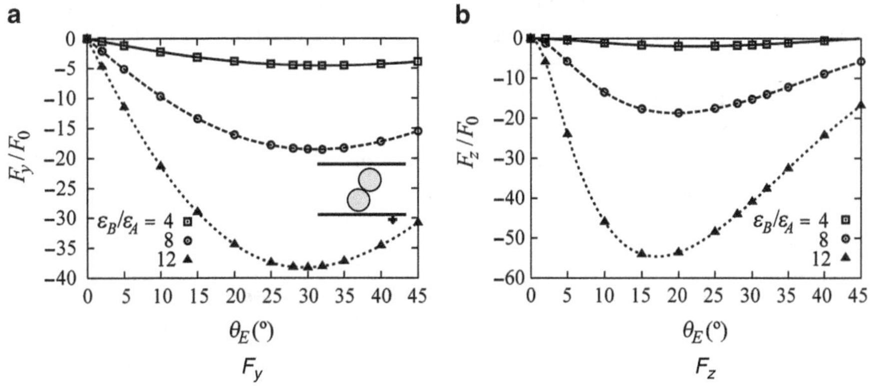

Fig. 6.20 Force components on the upper particle in Fig. 6.16b [21]

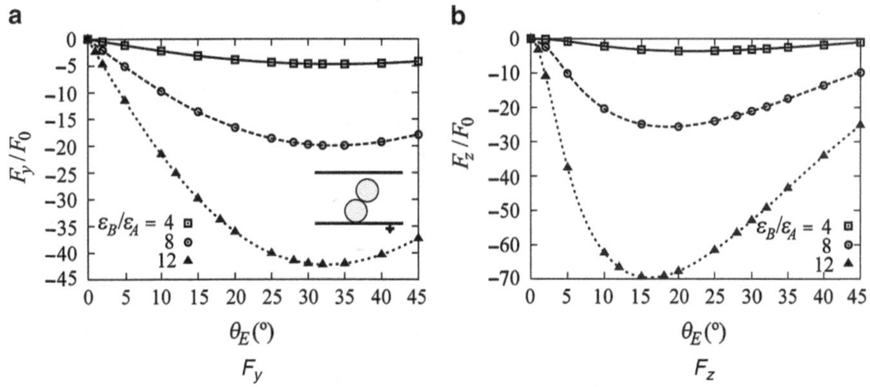

Fig. 6.21 Force components on the upper particle in Fig. 6.16c [21]

The horizontal force F_y is of particular interest here. Positive or negative F_y means that the DEP force reinforces or opposes the applied shear, respectively. In all the configurations considered here, Figs. 6.19–6.21 show that F_y is negative for θ_E up to 45°, i.e., F_y resists the shear. The absolute value $|F_y|$ increases from zero at zero θ_E, reaches a maximum at a certain critical angle, and then decreases with further increases in θ_E. From this characteristic behavior of F_y, we can conclude that with application of shear, a particle chain is tilted to angle θ_E at which the DEP force balances the shear. If this balancing does not occur by the time θ_E reaches the critical angle, then the chain is completely broken. The corresponding shear magnitude is called the shear yield stress [1], which can be determined by experiment. The maximum value of $|F_y|$ varies with the configuration. It is smallest for the chain of Fig. 6.16a, where the particles are separated from each other but are still in contact with the electrodes. The maximum value is similar for the other two configurations, where the particles are still in contact with each other, although the force is slightly stronger in Fig. 6.21 for the chain attached to the lower electrode.

The vertical force F_z shown in Figs. 6.19–6.21 illustrates that the upper particle is attracted to the upper plane electrode in the configuration of Fig. 6.16a, but is pulled downward in the other configurations. This implies that the interaction between the particle and the electrode is dominant for the chain shown in Fig. 6.16a, whereas the interaction between the particles is dominant for the other chain configurations. The magnitude of F_z does not vary so significantly among the three configurations.

6.4.4 Approximation of the Maximal Horizontal Force

When a chain of spherical particles is tilted with respect to the applied electric field, a three-dimensional (3D) field calculation is required to obtain information on the field and force, as shown in the preceding sections. Instead of performing such a complicated computation, an approximate calculation has been applied, based on a model of a two-particle chain by L.C. Davies [22] and P. Gonon et al. [23], or an infinitely-long particle chain by C.W. Wu and H. Conrad [24, 25]. Here, we shall explain and compare the approximate results with those from the accurate analysis.

For a chain making angle θ_E with the external field, the force is often estimated from F_{a0} for the case $\theta_E = 0°$ (see the left part of Fig. 6.22) on the assumption that only the component $E_0\cos(\theta_E)$, parallel to the chain, contributes to the force.

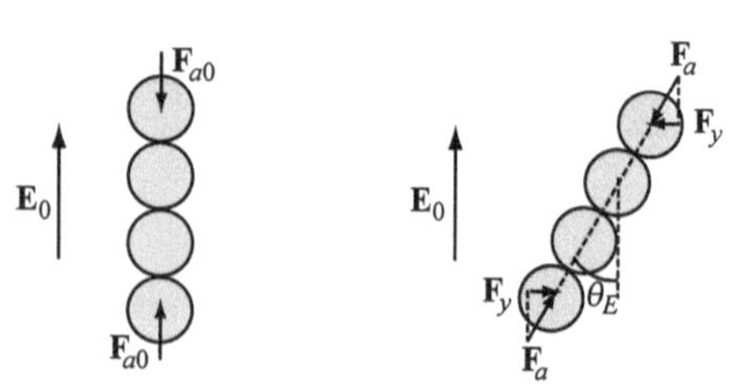

Fig. 6.22 Model of an isolated particle chain for the approximate calculation

Therefore, for arbitrary θ_E, the electric field E_θ and its normal component $E_{n\theta}$ on the particle surface can be related to the field $E_{\theta,0}$ and its normal component $E_{n\theta,0}$ in the case $\theta_E = 0°$ by

$$\frac{E_\theta}{E_{\theta,0}} = \frac{E_{n\theta}}{E_{n\theta,0}} = \cos\theta_E. \tag{6.7}$$

Integrating Maxwell's stress over the particle surface, as in Eq. 6.2, we obtain the force $\mathbf{F}_a$ in Fig. 6.22 for tilt angle θ_E as

$$\mathbf{F}_a = \varepsilon_A\varepsilon_0 \int \left(\mathbf{E}_\theta E_{n\theta} - \frac{1}{2} E_\theta^2 \mathbf{a}_n \right) ds, \tag{6.8}$$

where $\mathbf{a}_n$ is the unit normal vector on the surface. From Eqs. 6.7 and 6.8, it can be inferred that

$$F_a = F_{a0} \cos^2\theta_E. \tag{6.9}$$

In Eq. 6.9, F_{a0} is the force magnitude where $\theta_E = 0$. Hence, the horizontal force for a chain at angle θ_E is approximated as

$$F_y = F_{a0} \cos^2\theta_E \sin\theta_E. \tag{6.10}$$

The maximum of F_y is then determined as

$$\max\{F_y\} = F_{a0} \max\{\cos^2\theta_E \sin\theta_E\} = 0.385 F_{a0}, \tag{6.11}$$

which occurs when θ_E is 35.3°; this value is independent of $\varepsilon_B/\varepsilon_A$ although F_{a0} depends on $\varepsilon_B/\varepsilon_A$. In comparison, the accurate 3D analysis in this section shows that the critical angle is dependent on $\varepsilon_B/\varepsilon_A$ to a certain extent. Also, the angle at maximal F_y is slightly smaller than 35.3° for the chain shown in Figs. 6.16b and c, but it is noticeably smaller than the approximation if the particles are separated from each other as in the configuration of Fig. 6.16a.

The $\max\{F_y\}$ estimated by Eq. 6.11 is shown in Table 6.4 for the particle number $N = 2$, whereas the accurate numerical results for the configurations in Figs. 6.16a, b, and c are given in Table 6.5. Comparing the force magnitudes in the tables, we can see that for small $\varepsilon_B/\varepsilon_A$, the approximation of Eq. 6.11 for an isolated two-particle chain gives a value relatively close to that for the chain shown in Fig. 6.16a, but the approximation is smaller than the true value for the other configurations.

Table 6.4 Approximate values for $\max\{F_y\}/F_0$ on an isolated particle chain [21]

Particle number, N	$\varepsilon_B/\varepsilon_A$		
	4	8	12
2	2.45	9.96	20.4

Table 6.5 Accurate values for $\max\{F_y\}/F_0$ based on 3D numerical calculations [21]

Figure showing chain configuration	$\varepsilon_B/\varepsilon_A$		
	4	8	12
6.16a	2.38	7.53	12.8
6.16b	4.55	18.5	38.3
6.16c	4.70	19.9	42.3

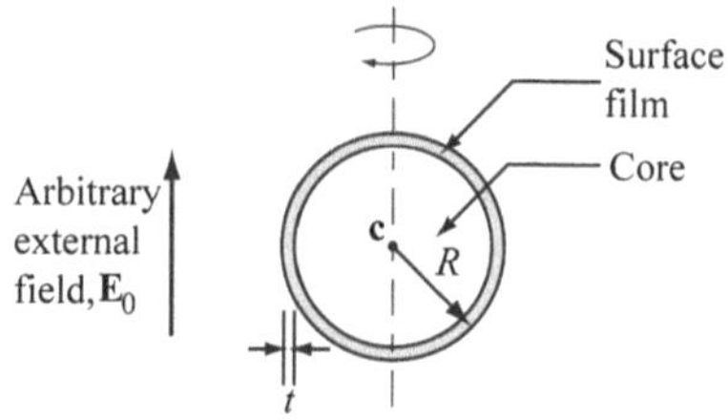

Fig. 6.23 Structure of a spherical particle with a surface film

6.5 Nonhomogeneous Particles

6.5.1 Particles with a Surface Film

Various types of particles have been proposed for improving the performance of ER fluids [22–25]. They may be classified into two main categories:

1. Homogeneous particles without any surface element, and
2. Particles composed of a core and a surface film (nonhomogeneous or layered particles).

Although more complicated forms such as multilayer particles are possible in theory, they are rarely found in practical use. The preceding sections of this chapter exclusively consider homogeneous spherical particles.

This section deals with two different types of layered spherical particles, i.e. (a) conductor-core particles and (b) dielectric-core particles. Figure 6.23 shows the structure of a particle composed of a core and a surface film having total radius R and film thickness t. The conductivity σ_C of the core is infinity for the former and zero for the latter. The surface film is assumed to have finite conductivity σ_F. The applied voltage is confined in this section to dc and low-frequency ac excitation in which the conductivity is predominant over the permittivity. For general cases, one may apply an analogous calculation method by replacing the conductivity with the complex term $\dot{\sigma} = \sigma + j\omega\varepsilon\varepsilon_0$.

An important concept applied to a layered particle is the apparent conductivity (or apparent permittivity), which is used to equivalently represent the effects that the particle exerts on the field outside the particle itself. This concept has been proposed for dielectric mixtures or a layered spherical particle existing in a uniform field by R.D. Stoy, A. Sihvola et al., and A.H. Sihvola et al. [15, 26, 27]. In a system of multiple particles, the presence of the particles themselves modifies the field experienced by each particle, and the response to a nonuniform field is important for understanding the electric field and force behavior.

6.5.2 Calculation Method

Based on the re-expansion method explained in Section 8.4, we can expand the external potential ϕ_0 for a particle due to all exterior sources in the following form:

$$\phi_0 = \sum_{j=0}^{\infty} \sum_{k=-j}^{j} M_{j,k} r_c^j \bar{P}_{j,|k|}(\cos\theta_c) \exp(\mathrm{j}k\varphi_c), \tag{6.12}$$

where $M_{j,k}$ is the potential coefficient, j is the imaginary unit, $\bar{P}_{j,|k|}$ is the normalized Legendre function, and $(r_c, \theta_c, \varphi_c)$ are spherical coordinates. The origin of the spherical coordinates $(r_c, \theta_c, \varphi_c)$ is taken at the center **c** of the particle, as indicated by the subscripts 'c', and θ_c is zero in the direction of the field at **c**. With the presence of a particle, the resultant potential ϕ_E outside the particle becomes

$$\phi_E = \sum_{j=0}^{\infty} \sum_{k=-j}^{j} \left(M_{j,k} r_c^j + \frac{B_{j,k}}{r_c^{j+1}} \right) \bar{P}_{j,|k|}(\cos\theta_c) \exp(\mathrm{j}k\varphi_c), \tag{6.13}$$

in which potential coefficients $B_{j,k}$ are the multipole potentials resulting from the surface charge on the particle. In particular, $B_{1,0}\cos\theta_c/r_c^2$ is the potential of a dipole with moment $p = 4\pi\varepsilon_A\varepsilon_0 B_{1,0}$. Therefore, if a particle is subjected to a uniform field, p will be the equivalent dipole moment induced on the particle by the field. The potential due to $B_{j,k}$ in Eq. 6.13 vanishes as r_c becomes infinitely large.

In a general case where the applied field is nonuniform, $B_{j,k}$ is also nonzero for orders (j, k) other than (1, 0). The boundary condition of current continuity on the particle surface results in the following relationship of potential coefficients:

$$B_{j,k} = M_{j,k} \frac{j(\sigma_B - \sigma_{app,j})}{(j+1)\sigma_B + j\sigma_{app,j}} R^{2j+1} \tag{6.14}$$

6.5.3 Apparent Conductivity

In Eq. 6.14, $\sigma_{app,j}$ is defined as the apparent conductivity of the particle under a j-th order external potential $\phi_{0,j,k} = M_{j,k} r_c^j \bar{P}_{j,|k|}(\cos\theta_c) \exp(\mathrm{j}k\varphi_c)$ for any k. For a homogeneous particle with conductivity σ_B, the apparent conductivity is simply expressed as

$$\sigma_{app,j} = \sigma_B, \tag{6.15}$$

independent of j and k.

For a particle composed of a conductor core and a surface film,

$$\sigma_{app,j} = \left[\frac{j + (j+1)\zeta^{2j+1}}{j(1 - \zeta^{2j+1})}\right]\sigma_F \quad \text{for} \quad j \geq 1, \tag{6.16}$$

and for a particle composed of a dielectric core and a surface film,

$$\sigma_{app,j} = \left[\frac{(j+1)(1 - \zeta^{2j+1})}{(j+1) + j\zeta^{2j+1}}\right]\sigma_F \quad \text{for} \quad j \geq 1, \tag{6.17}$$

In Eqs. 6.16 and 6.17, the apparent conductivity is a function of the film conductivity σ_F, the order j of the applied potential, and the geometrical parameter $\zeta = t/R$, the ratio of film thickness to particle radius. Although $\sigma_{app,j}$ changes neither with j nor with k for a homogeneous particle, it depends on j for a layered particle. This means that even if a homogeneous particle and a layered particle respond to a uniform field in exactly the same way, their response to a nonuniform field will be different. In other words, it is usually impossible to replace a layered particle with an appropriate homogeneous particle and obtain the same response to an applied electric field. The expression of $\sigma_{app,j}$ can also be derived for the general case in which the core is neither a conductor nor a dielectric but possesses a finite conductivity [28].

The variation of $\sigma_{app,j}$ with j is given as an example for three ratios of t/R in Fig. 6.24. For zero thickness, i.e., when no film exists at all, the conductivity is infinite for a conductor-core particle and zero for a dielectric-core particle. With increasing film thickness, $\sigma_{app,j}$ decreases for the conductor core but increases for the dielectric core. It must be equal to σ_F for $t/R = 1$, corresponding to the case of a homogeneous particle of conductivity σ_F. On a particle with a fixed t/R ratio, $\sigma_{app,j}$ decreases with j for a conductor core and increases for a dielectric core. Figures 6.24a and b also

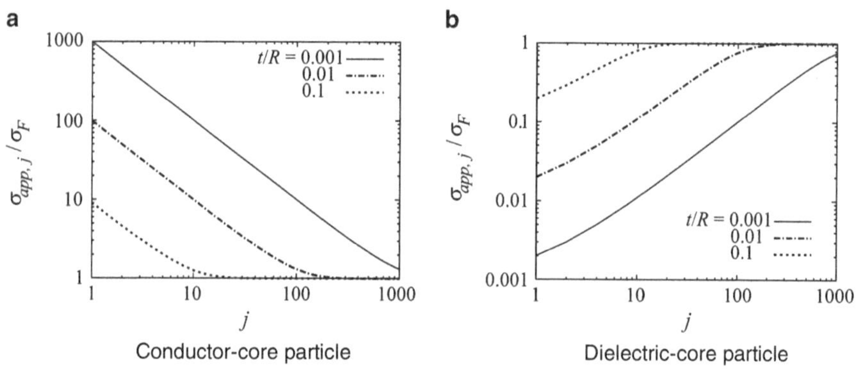

Fig. 6.24 Apparent conductivity with order j for three ratios of t/R [28]. © 2004 AIP

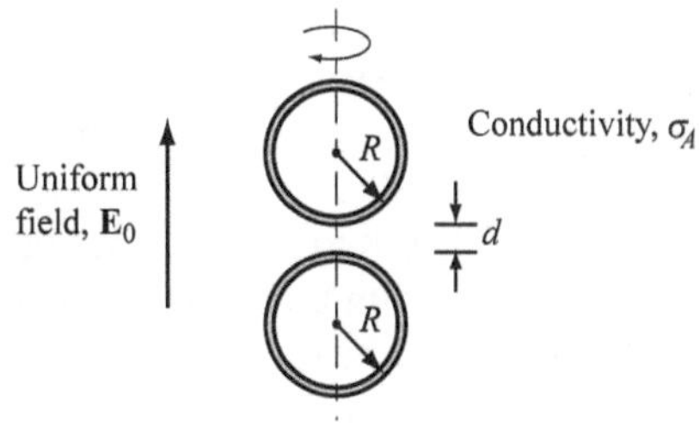

Fig. 6.25 Two particles separated by distance d under uniform field $\mathbf{E}_0$

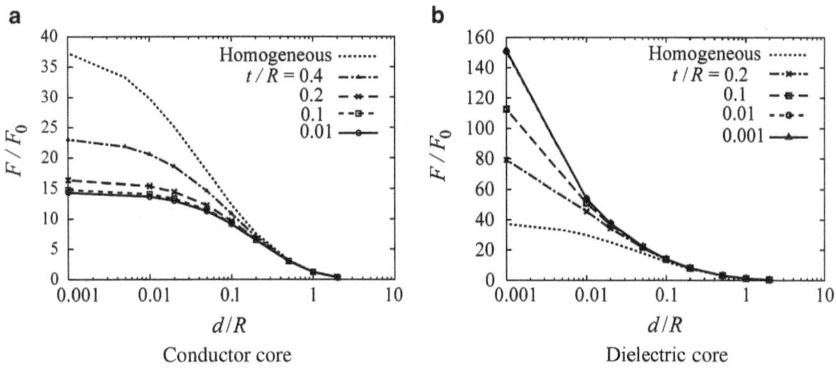

Fig. 6.26 Force on a particle in the configuration shown in Fig. 6.25 as a function of separation d [28]. © 2004 AIP

indicate that the surface film plays a more significant role on the response of a particle to a higher-order field, as $\sigma_{app,j}$ approaches σ_F with increasing order j.

The clear difference in the force behavior between a homogeneous particle and a layered one is seen, for example, in the configuration of two particles separated in a medium of conductivity σ_A under uniform field $\mathbf{E}_0$, as shown in Fig. 6.25. Figure 6.26 presents the attractive DEP force acting on the particles as a function of separation d for homogeneous particles without a surface layer and for four kinds of layered particles. In the calculation for homogeneous particles, the force corresponds to particle conductivity $\sigma_B = 10\sigma_A$. For the layered particles, surface conductivity σ_F is dependent on the film thickness and was chosen so that $\sigma_{app,1} = 10\sigma_A$. Thus, all the particles would exhibit the same interaction if they were adequately isolated under a uniform applied field.

It can be seen from Fig. 6.26 that the surface film decreases the force for conductor-core particles compared with the case of homogeneous particles, but the effect is reversed for dielectric-core particles. This difference between homogeneous and layered particles almost disappears when the separation is greater than the particle radius. This confirms the fact that these particles behave in the same manner as under a uniform field when they are far enough apart. As the particles get closer together, the external field that each particle experiences is modified, resulting in an external field of higher order. The presence of higher-order fields leads to the difference in force behavior from that of homogeneous particles. Since $\sigma_{app,j}$ is lower than $\sigma_{app,1}$ for a particle with a conductor core, the force magnitude due to higher-order fields is smaller than that on homogeneous particles, for which all $\sigma_{app,j}$ are of the same value. Thus, the force becomes stronger on conductor-core particles in Fig. 6.26a with increasing film thickness. Similar deductions can be made for particles with a dielectric core, which, however, exhibit the reversed tendency, as the apparent conductivity $\sigma_{app,j}$ ($j > 1$) is higher than $\sigma_{app,1}$.

References

1. Jordan, T.C., Shaw, M.T.: Electrorheology. IEEE Trans. Electr. Insul. **24**(5), 849–878 (1989)
2. Winslow, W.: Field responsive fluid couplings. US Patent No. 2, 886,151 (1959)
3. Winslow, W.: Induced fibration of suspensions. J. Appl. Phys. **20**(12), 1137–1140 (1949)
4. Bullough, W.A. (ed.): Proceedings of the 5th International Conference on ER Fluids, MR Suspensions and Associated Technology, World Scientific Publising, Singapore (1995)
5. Nakao, M., Koyama, K. (ed.): Proceedings of the 6th International Conference on ER Fluids, MR Suspensions and Associated Technology, World Scientific Publising, Singapore (1997)
6. Tao, R. (ed.): Proceedings of the 7th International Conference on ER Fluids and MR Suspensions, World Scientific Publising, Singapore (1999)
7. Bossis, G. (ed.): Proceedings of the 8th International Conference on ER Fluids and MR Suspensions, World Scientific Publising, Singapore (2001)
8. Gordaninejad, F., Graeve, O.A., Fuchs, A., York, D. (ed.): Proceedings of the 10th International Conference on ER Fluids and MR Suspensions. World Scientific Publising, Singapore (2006)
9. Yoshida, K., Kikuchi, M., Park, J.-H., Yokota, S.: Fabrication of micro electro-rheological valves by micromachining and experiments. Sensor. Actuat. A **95**(2–3), 227–233 (2002)
10. Pohl, H.A.: Dielectrophoresis. Cambridge University Press, Cambridge (1978)
11. Smythe, W.R.: Static and dynamic electricity, pp. 6–7. McGraw-Hill, New York (1968)
12. Davis, M.H.: Electrostatic field and force on a dielectric sphere near a conducting plane – a note on the application of electrostatic theory to water droplets. Am. J. Phys. **37**(1), 26–29 (1969)
13. Washizu, M., Jones, T.B.: Dielectrophoretic interaction of two spherical particles calculated by equivalent multipole-moment method. IEEE Trans. Ind. Appl. **32**(2), 233–242 (1996)
14. Stoy, R.D.: Induced multipole strengths for two dielectric spheres in an external electric field. J. Appl. Phys. **69**(5), 2800–2804 (1991)
15. Stoy, R.D.: Solution procedure for the Laplace equation in bispherical coordinates for two spheres in a uniform external field: parallel orientation. J. Appl. Phys. **65**(7), 2611–2615 (1989)
16. Jones, T.B.: Electromechanics of particles. Cambridge University Press, Cambridge (1995)
17. Nakajima, Y., Matsuyama, T.: Calculation of pearl chain forming force by re-expansion method. J. Inst. Electrost. Jpn. **25**(4), 215–221 (2000) (in Japanese)
18. Techaumnat, B., Takuma, T.: Calculation of the electric field for lined-up spherical dielectric particles. IEEE Trans. Dielect. Electr. Insul. **10**(4), 623–633 (2003)
19. Takuma, T.: Field behavior in composite dielectric arrangements. IEEE Trans. Electr. Insul. **26**(3), 500–509 (1991)
20. Techaumnat, B., Eua-arporn, B., Takuma, T.: Calculation of electric field and dielectrophoretic force on spherical particles in chain. J. Appl. Phys. **95**(3), 1586–1593 (2004)
21. Techaumnat, B., Eua-arporn, B., Takuma, T.: Electric field and dielectrophoretic force on a dielectric particle chain in a parallel-plate electrode system. J. Phys. D: Appl. Phys. **37**(23), 3337–3346 (2004)

22. Davis, L.C.: Time-dependent and nonlinear effects in electrorheological fluids. J. Appl. Phys. **81**(4), 1985–1990 (1997)
23. Gonon, P., et al.: Particle–particle interactions in electrorheological fluids based on surface conducting particles. J. Appl. Phys. **86**(12), 7160–7169 (1999)
24. Wu, C.W., Conrad, H.: Influence of a surface film on the particles on the electrorheological response. J. Appl. Phys. **81**(1), 383–389 (1999)
25. Wu, C.W., Conrad, H.: Influence of a surface film on conducting particles on the electrorheological response with alternating current fields. J. Appl. Phys. **81**(12), 8057–8063 (1999)
26. Sihvola, A., Lindell, I.V.: Effective permittivity in microwave remote sensing problems: media as mixtures with scatterers of Gaussian packets. Proc. 18th Eur. Microw. Conf., 693–698 (1988)
27. Sihvola, A.H., Kong, J.A.: Effective permittivity of dielectric mixtures. IEEE Trans. Geosci. Remote Sens. **26**(4), 420–429 (1988)
28. Techaumnat, B., Takuma, T.: Electric field and dielectrophoretic force on particles with a surface film. J. Appl. Phys. **96**(10), 5877–5885 (2004)

Chapter 7
Electric Field and Force on Toners in Electrophotography

Introduction

Charged particles are involved in many applications such as electrophotography, electrostatic painting, and electrostatic precipitators. Electrophotography was invented in 1938 as a copying technology by C.F. Carlson who named it xerography. Its commercialization began with the Xerox 914 copying machine by the Xerox Corporation (the Haloid Company at that time) about a half century ago. A historical review of the development and success of the corresponding industry is beyond the scope of this book.

Electrophotography generally includes various processes such as charging (of toners and photoconductors), exposure, and development. Among these processes, the electrostatic force on toner particles is of particular interest when the particles attach to an electrode or a photoconductor. Chapter 6 covers the behavior of the electric field and force for various configurations of uncharged particles. In these configurations, the particles are energized by an externally applied field, which results in the dielectrophoretic (DEP) force. This chapter describes the electrostatic force on a charged particle with and without an applied electric field, in particular, when the charge distribution on the particle is nonuniform.

The adhesion of microscopic charged particles is ascribed to van der Waals force and the electrostatic force. Although it is still in dispute which force plays a predominant role for toners in electrophotography [1, 2], we confine our discussion to the electrostatic force. The particles are treated as being spherical, as in the previous chapter.

T. Takuma and B. Techaumnat, *Electric Fields in Composite Dielectrics and their Applications*, Power Systems, DOI 10.1007/978-90-481-9392-9_7,

7.1 Fundamental Characteristics

7.1.1 Fundamentals of the Adhesive Force

Figure 7.1 shows a schematic of the configuration used in electrophotography. Toner particles are charged and positioned on a photoconductor. Without the applied field, the electrostatic force attracting toner particles to the photoconductor originates from the interaction of the toner charge with the photoconductor and the electrode on which it rests. In order to transfer the particles to another medium (a sheet of paper or an intermediate belt), electric field $\mathbf{E}_e$ is applied to overcome the adhesive force. However, the behavior of the force in the presence of $\mathbf{E}_e$ is complicated because the field also contributes to an additional DEP force enhancing the particle adhesion. The basic characteristics of the DEP force are explained in Chapter 6 for systems of uncharged particles.

The adhesion between toner particles and the photoconductive substrate (photoconductor) in the absence of $\mathbf{E}_e$ is of particular interest because it can be measured by using techniques such as the measurement of centrifugal force [3, 4]. To analytically estimate the adhesive force F_a, a simple configuration of a spherical dielectric particle lying on a conducting plane is often employed. Assuming a uniform distribution of total charge Q on the particle surface, the force is given by

$$F_a = \frac{Q^2}{16\pi\varepsilon_0 R^2}, \tag{7.1}$$

where R is the particle radius. In this chapter, the dielectric constant ε_A (relative permittivity) of the surrounding medium is equal to unity, as air is the typical medium in electrophotography. This equation excludes the contribution from particle polarization, which results in additional, nonuniform surface charges (polarization charge), unless the dielectric constant ε_B of the particle is equal to ε_A.

Experimental measurements report much stronger electrostatic adhesion (by one or two orders of magnitude) than the approximate value given by Eq. 7.1 [5]. More

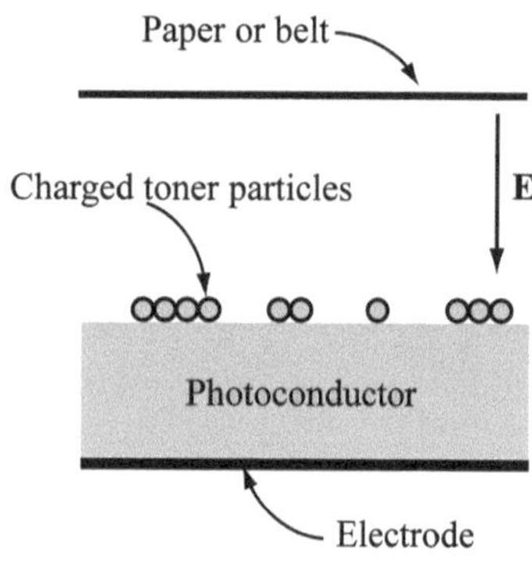

Fig. 7.1 Charged toner particles subjected to applied field $\mathbf{E}_e$

accurate analytical studies that include the effect of polarization show that the force on a uniformly charged particle increases with ε_B of the particle [6]. However, for $\varepsilon_B = 3$–4, which is typical for toners, the analytical force is still much smaller than the measured values. Another possible cause for the underestimation when applying Eq. 7.1 may be ascribed to the assumption of a uniform charge distribution on the toner surface. Unlike a conducting particle, where charge is distributed to fulfill the condition of a constant potential, charge distribution on a dielectric toner particle depends on the charging process, which is usually done via tribology. Some types of nonuniform charge distributions, such as the patch type and dumbbell type, have already been analyzed [5, 7–10]. However, the resulting forces were estimated by empirical expressions or numerical computations of relatively low resolution.

7.1.2 Nonuniform Charging Models

Two kinds of nonuniform charge distribution are used here to elucidate the effect on the electrostatic force. The surface charge density q_s on a particle is defined as a function of the zenith angle θ, as shown in Fig. 7.2. For simplicity, q_s is assumed to be independent of the azimuth angle φ, i.e., the system is rotationally symmetrical. Since the Coulomb force acts so as to attract the charged site on a particle to the photoconductor (see Fig. 7.1), it is likely that the most highly charged region will be brought into contact with the photoconductor.

For the first kind of distribution, the charge density is defined as

$$q_s = \varepsilon_0 K_q \{1 - \cos[k_q \theta - (k_q - 1)\pi]\} \tag{7.2}$$

for $\theta = (k_q - 1)\pi/k_q$ to π, where k_q is an integer, and $q_s = 0$ on the remaining portion of the surface. The density is maximal at the bottom pole and smoothly declines to zero with decreasing θ. The degree of nonuniformity is varied by changing the range of θ $(= \pi/k_q)$ on which the charge exists, i.e., higher k_q values give a smaller charged region and a higher degree of charging nonuniformity. In Eq. 7.2, K_q is a real number parameter used to adjust the total charge on a particle. An example of the charge distribution with θ is given in Fig. 7.3a for some values of k_q ranging from 1 to 8.

The second kind of distribution is the so-called dumbbell type. This distribution is based on the suggestion that triboelectrification may result in a dumbbell-shaped

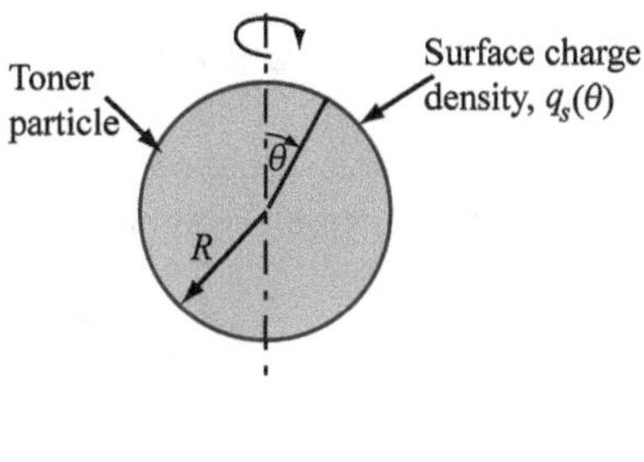

Fig. 7.2 Surface charge density q_s defined as a function of the zenith angle θ

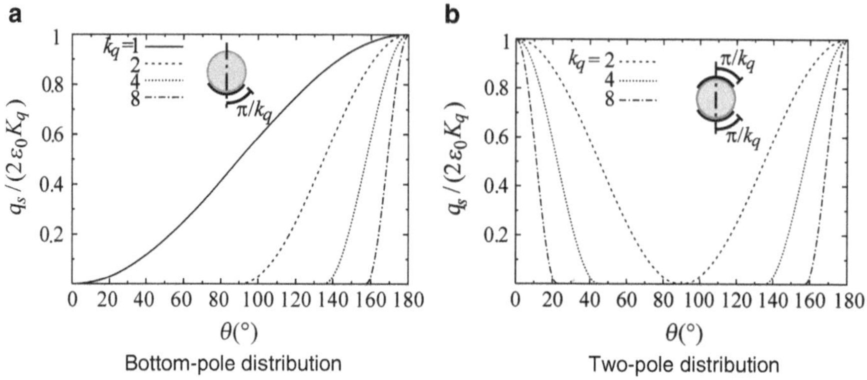

Fig. 7.3 Two kinds of charge distribution

charge distribution [10]. In addition to the charge at the bottom pole, there is another charge site at θ ranging from 0 to π/k_q with the density.

$$q_s = \varepsilon_0 K_q\{1 + \cos(k_q\theta)\} \tag{7.3}$$

This kind of charge distribution has a vertical symmetry with respect to the plane $\theta = \pi/2$, as shown in Fig. 7.3b.

In the following analysis, we shall refer to the first kind of charge distribution as the bottom-pole distribution and the second kind as the two-pole distribution. Unless otherwise indicated in the text, comparison between these two cases shall be made for the same total charge. Although these charging models do not exactly represent the actual charge distribution on the particle, they are used to study the force behavior in two fundamental cases. The bottom-pole distribution represents the case in which a large amount of charge is concentrated on a portion of the particle surface. On the other hand, the two-pole distribution is used for observing the variation of force when there is another highly charged site located far from the first one.

7.1.3 Field Calculation

The re-expansion method (method of multipole images) described in Section 8.4 has been used for calculating the electric field for various configurations of charged particles. Here we will explain only the main concepts directly concerning the charge distribution defined by Eq. 7.2 or 7.3.

In the re-expansion method, the potentials ϕ^I internal and ϕ^E external to the particle are expressed separately as the infinite series of spherical harmonics as

$$\phi^I = \sum_{j=0}^{\infty} L_j r^j P_j(\cos\theta) \tag{7.4}$$

$$\phi^E = \sum_{j=0}^{\infty} \left(M_j r^j + \frac{B_j}{r^{j+1}} \right) P_j(\cos\theta) \tag{7.5}$$

where the origin of the coordinates (r, θ) is at the particle center, and P_j is the Legendre function. The azimuth angle can be excluded from the expressions for ϕ^I and ϕ^E due to the axisymmetry of the configuration under consideration. Equation 7.5 is valid for r ranging from R to an appropriate radius depending on the configuration. The potential coefficients L_j, M_j, and B_j can be determined uniquely to satisfy the following boundary conditions.

On the conducting plane,

$$\phi^E = V_0(\text{constant}) \tag{7.6}$$

On the particle surface ($r = R$),

$$\phi^E = \phi^I, \tag{7.7}$$

and

$$\varepsilon_B \frac{\partial}{\partial r} \phi^I - \frac{\partial}{\partial r} \phi^E = \frac{q_s}{\varepsilon_0}. \tag{7.8}$$

To apply the condition of Eq. 7.8 to the calculation, the charge density q_s is interpolated with the Legendre function as

$$\frac{q_s}{\varepsilon_0} = \sum_{j=0}^{\infty} S_j P_j(\cos\theta). \tag{7.9}$$

The coefficient S_j can be determined numerically by integrating

$$S_j = \frac{2n+1}{2} \int_{-1}^{1} \frac{q_s}{\varepsilon_0} P_j(\cos\theta) \mathrm{d}\cos\theta. \tag{7.10}$$

With the charge density expressed by Eq. 7.9 and the boundary conditions in Eqs. 7.7 and 7.8, $M_j = 0$ in the absence of E_e, and we can relate L_j and B_j to S_j as follows:

$$L_j = S_j \frac{R^{-j+1}}{j\varepsilon_B + (j+1)}, \tag{7.11}$$

$$B_j = S_j \frac{R^{j+2}}{j\varepsilon_B + (j+1)}. \tag{7.12}$$

On the other hand, with the presence of uniform field E_e but without surface charge, the corresponding potential coefficients that fulfill Eqs. 7.7 and 7.8 are

$$M_1 = -E_e, \tag{7.13}$$

$$L_1 = -E_e \frac{3}{\varepsilon_B + 2}, \tag{7.14}$$

$$B_1 = E_e R^3 \frac{\varepsilon_B - 1}{\varepsilon_B + 2}, \tag{7.15}$$

and M_j, L_j, and $B_j = 0$ for $j > 1$. In this case, $B_0 = 0$; and L_0 and M_0 are the constant components of the potentials, which do not contribute to the field strength.

Calculation of the electric field for a configuration of charged particles begins with a combination of the potentials in Eqs. 7.11–7.15. This initial solution satisfies the boundary conditions of an isolated, charged dielectric sphere under E_e without the conducting plane. Then, an image scheme with respect to the conducting plane and the re-expansion formulae described in Section 8.4 can be applied repetitively until the potential coefficients converge. The electrostatic force **F** on a particle is then determined from the integral of the Maxwell's stress over the particle surface, as explained in Chapter 6.

7.2 Charged Dielectric Particle on a Conductor

7.2.1 General Expression of Electrostatic Force

In this section, we consider the electrostatic force in the configuration shown in Fig. 7.4 where a charged dielectric sphere (particle) lies on a conducting plane under an externally applied electric field. This arrangement corresponds to the special case of a photoconductor of infinitesimally small thickness; thus the particle charge does not leak to the conducting plane. We may assume without loss of generality that the particle is positively charged and the applied field $\mathbf{E}_e$ is in the upward direction for the purpose of lifting the particle from the plane. The dielectric constant is equal to unity in the surrounding medium, as already explained.

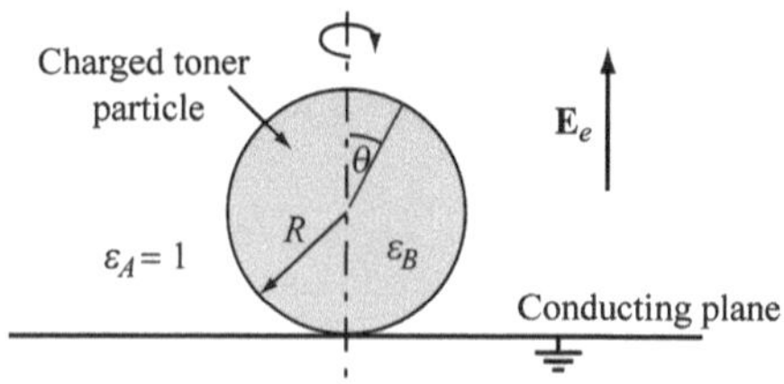

Fig. 7.4 A charged dielectric sphere lying on a grounded plane under applied electric field $\mathbf{E}_e$

In general, the electrostatic force on the particle shown in Fig. 7.4 can be separated into the three terms, Q^2, QE_e, and $E_e^{\,2}$, as described by J.Q. Feng and D.A. Hays [10]

$$F_a = \alpha \frac{Q^2}{R^2} - \beta Q E_e + \gamma E_e^2 R^2, \tag{7.16}$$

where Q is the total charge on the particle, R is the particle radius, and the coefficients (α, β, and γ) depend on the dielectric constants of the media involved and the distribution of charge. The adhesive force F_a is in the downward direction, attracting the particle to the plane. Thus, negative F_a lifts the particle from the plane. We may use a form slightly different from Eq. 7.16 for the force expression:

$$F_a = c_1 (V_Q - c_2 V_E)(V_Q - c_3 V_E), \tag{7.17}$$

where V_Q and V_E are the potentials corresponding to the total charge and the applied field, respectively. They are defined as

$$V_Q = \frac{Q}{4\pi\varepsilon_0 R}, \tag{7.18}$$

and

$$V_E = E_e R. \tag{7.19}$$

In Eq. 7.17, c_1 is the force per $V_Q^{\,2}$ in the absence of $\mathbf{E}_e$, while $c_2^{\,-1}$ and $c_3^{\,-1}$ are the ratios of V_E/V_Q that produce zero force on the particle. The quadratic form of Eq. 7.17 indicates that the force F_a is adhesive (positive) at low and high values of V_E. At low V_E, adhesion occurs as a result of the dominating interaction between the toner charge and the conducting plane below (in practice, the substrate). At sufficiently high V_E, polarization induced by the applied electric field is so high that the force due to the attraction between the polarization charge and its images is predominant. Hence, in order to lift a particle from the plane, we must apply an intermediate level of field between the lower and the upper limits defined by the two zeroes of F_a. That is to say, the Coulomb force QE_e prevails over the other interactions only in the intermediate range of E_e.

7.2.2 Adhesion in the Absence of an External Field

Figure 7.5 presents the adhesive force F_a on a particle with a bottom-pole charge distribution as a function of the dielectric constant ε_B for various values of k_q. Without an applied field, the force expression is reduced to $F_a = c_1 V_Q^{\,2}$. Thus, the normalized force $F_a/V_Q^{\,2}$ on the ordinates in Fig. 7.5 is equal to constant c_1 for each combination of ε_B and k_q, and can be used for comparing the force on particles having the same total charge (or V_Q). Figures 7.5a and b relate to small and large k_q values, respectively. The case of a uniform charge distribution is given as solid lines without intervening marks.

From Fig. 7.5, we can see that the concentration of charge near the bottom pole yields a substantial enhancement in adhesion because F_a increases with increasing k_q for each value of ε_B. For example, the force is about 55 times stronger for $\varepsilon_B = 3$ and $k_q = 8$ than that for a uniform charge distribution with the same ε_B. The effect of ε_B is rather complicated, as shown in the figures. At a low degree of charge nonuniformity (small k_q), the adhesion gets stronger with increasing ε_B, but this tendency is reversed for k_q larger than about 6.

The corresponding result for a particle with a two-pole charge distribution is shown in Fig. 7.6. The enhancement of the adhesive force with increasing k_q is also clearly observed, just as it was for the bottom-pole distribution, but to a lesser extent. However, on a particle with $\varepsilon_B = 3$, for example, the force at $k_q = 8$ is still approximately 15 times stronger than that for a uniform charge distribution. Increases in ε_B consistently result in a stronger adhesive force in the case of the two-pole charge distribution for k_q up to 8.

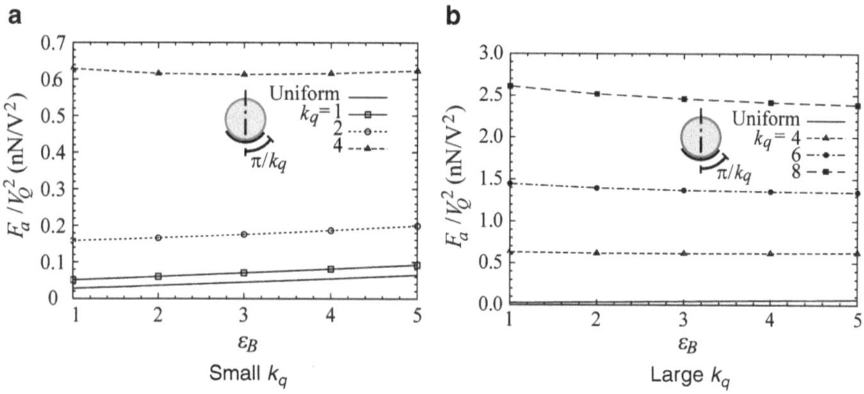

Fig. 7.5 Normalized adhesive force without applied field $\mathbf{E}_e$ on a particle with a bottom-pole distribution of charge [11]. © 2009 IEEE

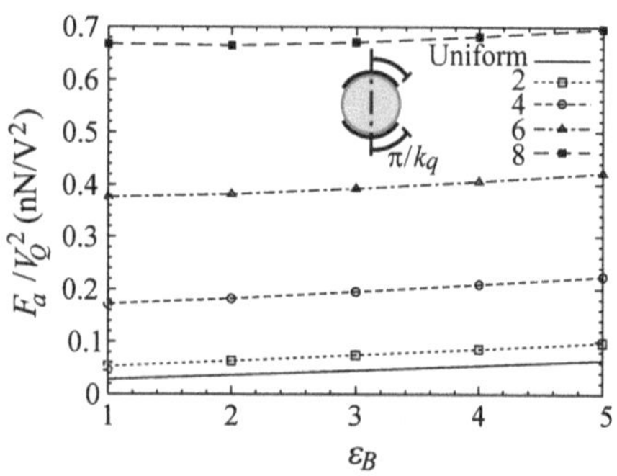

Fig. 7.6 Normalized adhesive force without $\mathbf{E}_e$ on a particle with a two-pole distribution of charge [11]. © 2009 IEEE

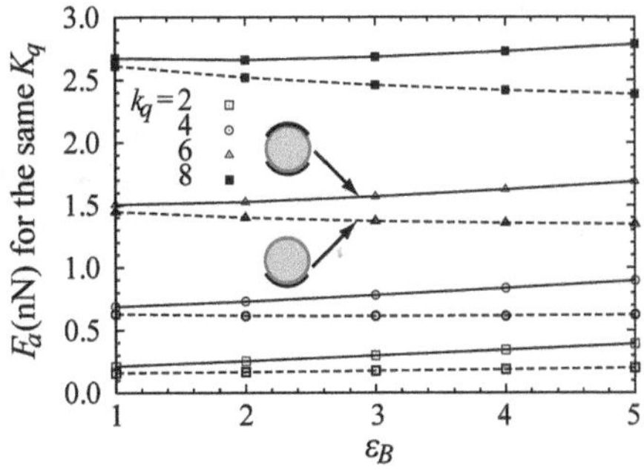

Fig. 7.7 Comparison of the adhesive force without $\mathbf{E}_e$ between the bottom-pole (*dashed lines*) and the two-pole distribution (*solid lines*) for the same value of K_q [11]. © 2009 IEEE

Figure 7.7 compares the adhesive force for the two kinds of charge distribution with the same value of K_q, i.e., the total charge Q and V_Q of the particle with the two-pole distribution is twice that for the particle with the bottom-pole distribution. In the figure, $V_Q = 1$ and 2V for the bottom-pole distribution and the two-pole distribution, respectively. Comparing the dashed line (corresponding to the bottom-pole distribution) with the solid line (two-pole distribution) for the same k_q, we can see that the additional charge located on the top pole contributes less in percentage terms to the total force F_a as k_q increases. However, the contribution becomes more important for higher ε_B of the particle.

7.2.3 Discrete Charge Distribution

Another model adopted to represent nonuniform charging is the discrete charge distribution with nonzero, constant q_s at the bottom pole or at the top pole of a particle, i.e.,

(a) Bottom-pole distribution, $q_s = q_{s0}$ from $\theta = (k_q - 1)\pi/k_q$ to π and $q_s = 0$ otherwise
(b) Top-pole distribution, $q_s = q_{s0}$ from $\theta = 0$ to π/k_q and $q_s = 0$ otherwise [12]

It should be noted that the meaning of k_q is somewhat different from that used in Eqs. 7.2 and 7.3 in Section 7.1.2. In this section, $k_q = 1$ means a uniform charge distribution, whereas $k_q = 1$ gives a spatially variable distribution in Eqs. 7.2 and 7.3.

In a different approach from the method employed for the charge distributions of Eqs. 7.2 and 7.3, the boundary element method (BEM) was used for calculating the field as it can more conveniently handle a discrete charge distribution. In this case, the normal components E_n of the electric field at the charged surface are related by

$$\varepsilon_A E_{nA} - \varepsilon_B E_{nB} = \frac{q_s}{\varepsilon_0}, \tag{7.20}$$

where the normal direction is defined in the direction outward from the particle, and the subscripts A and B denote the regions exterior and interior to the particle, respectively.

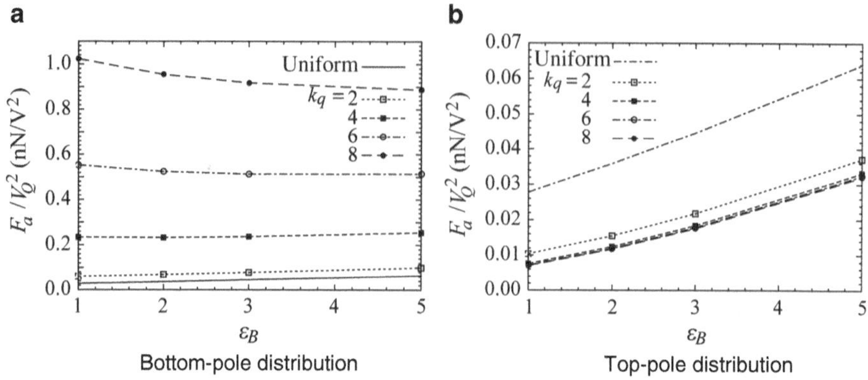

Fig. 7.8 Normalized adhesive force without applied field $\mathbf{E}_e$ on a particle with discrete charge distribution [12]. © 2009 Elsevier B.V.

The behavior of the adhesive force on the particle in this case is similar in qualitative terms to that explained in Section 7.2.2 for a continuous charge density. For example, Figs. 7.8a and b, respectively, show the electrostatic force on a particle with discrete charge on the bottom pole and that on a particle with discrete charge on the top pole. We can see from Fig. 7.8a that the force is stronger with charge at the bottom pole than on a uniformly charged particle and that the force becomes stronger with increasing k_q, in a similar way to that of the continuous charging model in Section 7.2.2. The reduction of the force with increasing ε_B is also observed for higher values of k_q. On the other hand, Fig. 7.8b clearly shows that the force is weaker than on a uniformly charged particle if charge is concentrated at the top pole. In this case, the reduction of the force with increasing ε_B is not seen.

In the following discussion, we again explain the results for the continuous charge distribution explained in Section 7.1.2.

7.2.4 Electrostatic Force Versus V_E

In the presence of an external field, the expression for F_a takes the full form of Eq. 7.17. The main focus here is on the field strength E_e that yields negative F_a, i.e., that lifts particles from the conducting plane. For the bottom-pole distribution of charge, an example of F_a for $\varepsilon_B = 3$ is displayed in Fig. 7.9a as a function of V_E $(= E_e R)$. In the figure, the calculation results are presented as symbols, whereas the lines are drawn by applying Eq. 7.17. The constant c_1 in Eq. 7.17 is obtained from the calculated force at zero E_e, while c_2 and c_3 are determined from the zeroes of F_a. The corresponding results for the two-pole distribution are displayed in Fig. 7.9b. It is clear from these figures that the numerically calculated values agree perfectly well with the lines given by Eq. 7.17.

Figures 7.9a and b confirm the quadratic nature of the electrostatic force and illustrate the fact that a particle can be lifted off the plane only by applying

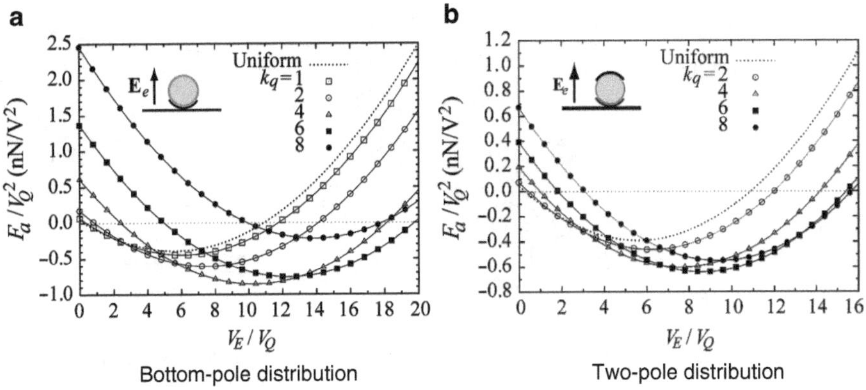

Fig. 7.9 Electrostatic force on a charged particle in the presence of $\mathbf{E}_e$ for $\varepsilon_B = 3$ (note that $V_E = E_eR$) [11]. © 2009 IEEE

intermediate levels of E_e to achieve negative F_a. This range of E_e for detachment depends highly on k_q, the degree of charging nonuniformity, especially for particles with the bottom-pole distribution. In addition to the change in the range of E_e for detachment, the maximum of the detachment force (negative peak in Fig. 7.9) is also highly dependent on k_q. In particular, we can see from Fig. 7.9a that the negative peak is significantly reduced for $k_q = 8$ for the bottom-pole charge distribution. For the two-pole distribution of charge shown in Fig. 7.9b, the maximum of the detachment force is stronger than that on a uniformly charged particle for all the values of k_q used here.

7.2.5 V_E for Detachment

The minimum and maximum values of V_E for detachment are given in Fig. 7.10 for $\varepsilon_B = 3$ (solid lines) and $\varepsilon_B = 4$ (dashed lines). The strongest lifting force is achieved at a middle point between the minimum and maximum. It can be seen from the figure that the minimum detachment V_E (and consequently E_e) becomes higher with larger k_q, i.e., with an increasing degree of charge nonuniformity. Since the bottom-pole distribution exhibits stronger adhesion than the two-pole distribution, it is expected that a higher E_e will be required to overcome the adhesion in the former case. On the other hand, with increasing k_q, the upper limit (maximum) of E_e becomes higher at small k_q but lower at large k_q.

The range of E_e values for detachment is markedly narrowed with increasing k_q for the bottom-pole distribution as shown in Fig. 7.10a, particularly in the case $k_q = 8$ and $\varepsilon_B = 4$. For the two-pole distribution, such critical reduction of the range of detachment E_e is not seen in Fig. 7.10b. Both figures clearly show that an increase in ε_B significantly decreases the maximum value of E_e, thus narrowing the ranges for detachment.

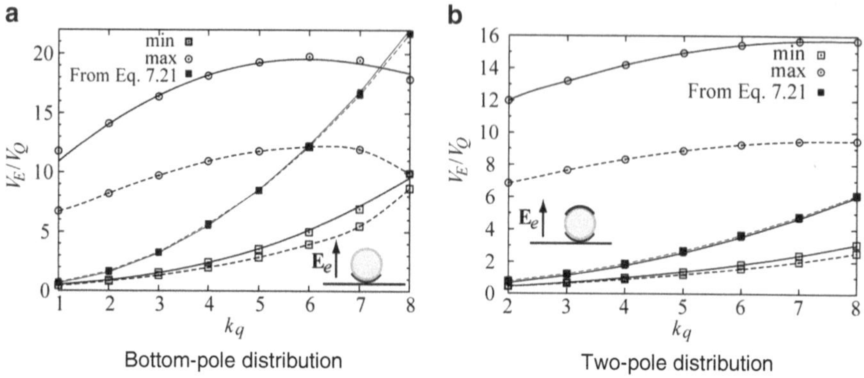

Fig. 7.10 V_E for detachment for $\varepsilon_B = 3$ (*solid lines*) and $\varepsilon_B = 4$ (*dashed lines*) [11]. © 2009 IEEE

The electric field needed for detachment may be roughly estimated from F_{a0}, the adhesive force when the external field does not exist, as the force acting on an isolated particle having charge Q,

$$QE_e = F_{a0}. \tag{7.21}$$

This estimation is compared with the analytical results in Figs. 7.10a and b. Clearly, E_e estimated by Eq. 7.21 is in all cases higher than the minimum field strength for detachment. In particular, the estimation may even exceed the upper limit of E_e for the bottom-pole distribution with high k_q. On the other hand, for the two-pole distribution, the estimation lies well within the minimum and maximum values obtained from the analysis.

7.3 Charged Dielectric Particle on a Dielectric Barrier

7.3.1 Configuration for Study

This section deals with the configuration of a particle lying on a dielectric barrier, which corresponds to the photoconductive material in electrophotography, as shown in Fig. 7.11. When an insulating solid exists between the particle and the conducting plane, the force behavior is more complicated as a result of additional interactions with the barrier. The force expression of Eq. 7.17 is still valid in the presence of the barrier; however, the constants c_1 to c_3 in this case depend also on the electrical properties and the thickness of the barrier. The focus of this section lies on the force of adhesion and detachment in relation with the barrier thickness, t. As the dielectric constant ε_S of the photoconductive material is not so different from that of the toner particles in practical applications, $\varepsilon_S = \varepsilon_B = 3$ is assumed in the analysis.

7.3.2 Adhesion in the Absence of an External Field

Figure 7.12 shows the normalized value of F_a versus t/R in the absence of E_e on a particle with the bottom-pole distribution. To clearly see the force behavior, the force is given for two overlapping ranges of t/R in parts (a) and (b). F_a at the limit of zero t/R in Fig. 7.12a is the adhesive force already described in the preceding section. It can be clearly seen from the figure that the presence of a dielectric barrier reduces the particle adhesion as F_a decreases with increasing t/R. When t is greater than about $2R$, the role of the conducting plane below the dielectric barrier virtually vanishes and the lower limit of F_a is reached, i.e., the adhesion results merely from the interaction between the particle and the dielectric barrier. The lower limit of F_a is significantly smaller than the upper limit at $t = 0$. However, the force enhancement that results from the nonuniform charge distribution still yields significantly stronger adhesion than that of a uniformly charged particle, which is shown with dashed lines in Fig. 7.12. For example, F_a/V_Q^2 is about 0.02 nN/V^2 when $t >> R$ for the uniform distribution of charge, but becomes 32 times higher for the bottom-pole distribution of charge with $k_q = 8$.

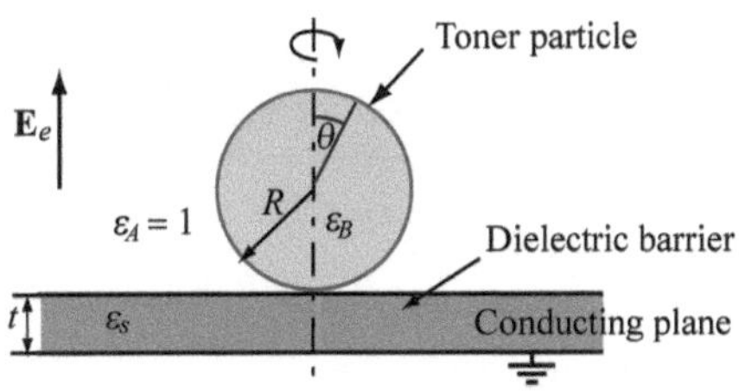

Fig. 7.11 Configuration of a charged particle on a dielectric barrier

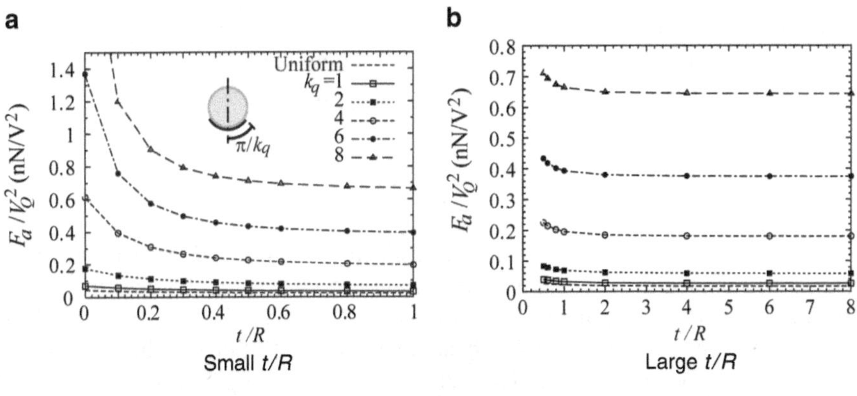

Fig. 7.12 Adhesive force versus barrier thickness t for $\varepsilon_B = \varepsilon_S = 3$ on a particle with bottom-pole distribution of charge

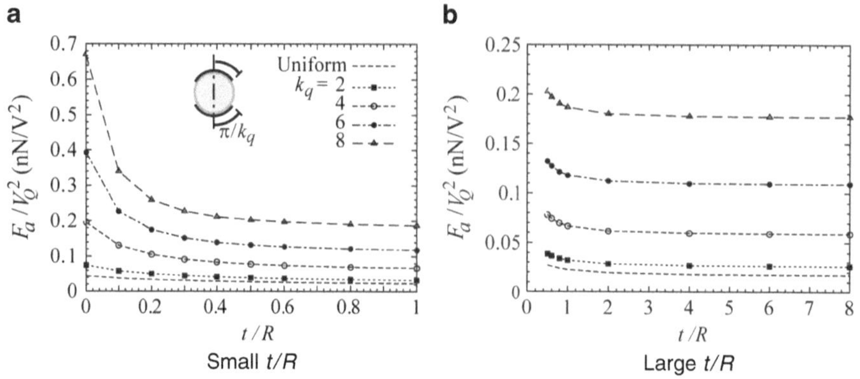

Fig. 7.13 Adhesive force versus barrier thickness t for $\varepsilon_B = \varepsilon_S = 3$ on a particle with two- pole distribution

Fig. 7.14 Lower limit of the adhesive force when $t >> R$ as a function of k_q for both kinds of charge distribution

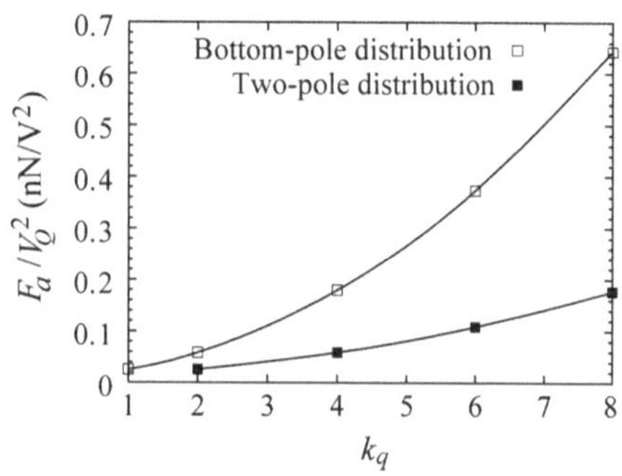

The adhesive force for the two-pole distribution is shown in Fig. 7.13; the force exhibits similar behavior to that in Fig. 7.12 but with a smaller force magnitude. For thickness t greater than $2R$, F_a is already close to the lower limit. For $k_q = 8$, the lower limit of F_a is still about ten times that on a uniformly charged particle. Figure 7.14 summarizes the lower limit of F_a for both kinds of charge distribution, and it clearly shows the nonlinear increase of F_a with k_q.

7.3.3 Detachment by an Applied Field

The level of V_E (= E_eR) required for detachment in the limit $t/R >> 1$ is shown in Fig. 7.15. In the figure, the maximum and minimum values of V_E for detachment are given as a function of k_q, together with the approximation given by Eq. 7.21. It is clear that changing k_q causes no significant reduction in the range of the detachment field either for the bottom-pole distribution or for the two-pole distribution. This characteristic is different from that found in the limit at $t/R = 0$, i.e., without a dielectric barrier. The maximal detachment force, depending on k_q, occurs at V_E

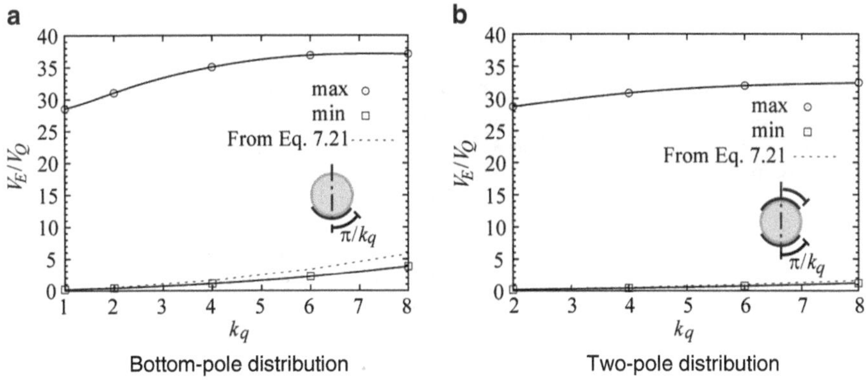

Fig. 7.15 Maximum and minimum of V_E for detachment where $t/R >> 1$

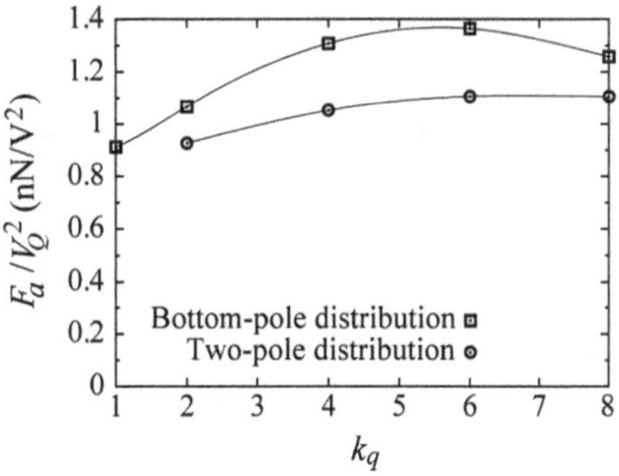

Fig. 7.16 Maximum of detachment force obtained by applying the optimal E_e in the limit $t/R >> 1$

between $15V_Q$ and $18V_Q$. V_E greater than about $35V_Q$ would fail to detach the particle.

Figure 7.16 displays the maximal detachment force for the two kinds of nonuniform charge distribution. The force is stronger where charge is concentrated near the bottom pole than near two poles. Increasing k_q results in a slightly stronger detachment force for the two-pole distribution. For the bottom-pole distribution, on the other hand, the maximal detachment force decreases when the distribution becomes highly nonuniform with k_q larger than about 6.

References

1. Rimai, D.S., Ezenyilimba, M., Goebel, W.K., Cormier, S., Quesnel, D.J.: Toner adhesion: effects of electrostatic and van der Waals interactions. J. Imaging Sci. Technol. **46**(3), 200–207 (2002)
2. Lee, M.: Comments on “Adhesion of silica-coated toner particles to bisphenol-A polycarbonate films”. J. Imaging Sci. Technol. **52**(2): 020103 (2008)

3. Dejesus, M.C., Rimai, D.S., Stelter, E., Tombs, T.N., Weiss, D.S.: Adhesion of silica-coated toner particles to bisphenol-A polycarbonate films. J. Imaging Sci. Technol. **52**(1): 010503 (2008)
4. Iimura, H., Kurosu, H., Yamaguchi, T.: Effects of an external additive on toner adhesion. J. Imaging Sci. Technol. **44**(5), 457–461 (2000)
5. Hays, D.A.: Toner adhesion. J. Adhes. **51**(1), 41–48 (1995)
6. Davis, M.H.: Electrostatic field and force on a dielectric sphere near a conducting plane – a note on the application of electrostatic theory to water droplets. Am. J. Phys. **37**(1), 26–29 (1966)
7. Fowlkes, W.Y., Robinson, K.S.: The electrostatic force on a dielectric sphere resting on a conducting substrate. In: Mittal, K.L. (ed.) Particles on surfaces 1: detection, adhesion and removal, pp. 143–155. Plenum, New York (1988)
8. Hays, D.A.: Electric field detachment of charged particles. In: Mittal, K.L. (ed.) Particles on surfaces 1: detection, adhesion and removal, pp. 351–360. Plenum, New York (1988)
9. Feng, J.Q., Eklund, E.A., Hays, D.A.: Electric field detachment of a nonuniformly charged dielectric sphere on a dielectric coated electrode. J. Electrost. **40–41**, 289–294 (1997)
10. Feng, J.Q., Hays, D.A.: Theory of electric field detachment of charged toner particles in electrophotography. J. Imaging Sci. Technol. **44**(1), 19–25 (2000)
11. Techaumnat, B., Kadonaga, M., Takuma, T.: Analysis of electrostatic adhesion and detachment of a nonuniformly charged particle on a conducting plane. IEEE Trans. Dielectr. Electr. Insul. **16**(3), 704–710 (2009)
12. Techaumnat, B., Takuma, T.: Analysis of the electrostatic force on a dielectric particle with partial charge distribution. J. Electrost. **67**(4), 686–690 (2009)

Chapter 8
Analytical Calculation Methods

Introduction

Although numerical calculation methods for solving Laplace's equation have developed remarkably in the past few decades, analytical methods still have significance in obtaining precise values and in easily evaluating parameter dependence, for example. Textbooks of electromagnetism explain a variety of analytical methods that were developed by great scientists and physicists, including J.C. Maxwell, in the nineteenth century. Among these methods, we confine our explanation in this chapter to some typical methods which have been used to calculate electric fields in composite dielectrics in this book. They are the variable-separation method for cross-sectionally straight dielectric interfaces, the iterative image charge method for a conducting sphere or cylinder on or above a solid dielectric plane, and the re-expansion method for configurations composed of spherical and planar boundaries.

8.1 Variable-Separation Method for Straight Dielectric Interfaces

The variable-separation method is applied for computing the electric field in configurations that have boundaries which are expressed with only one of the coordinates, with the other(s) being constant. Typical examples are monodielectric configurations in which conductors (electrodes) consist of straight equipotential surfaces meeting at one point. The edge point corresponds to the origin of the x-y or r-z coordinates. The configurations represent wedge-like edges, cone-pointed needles, or recesses of conductors. In this section, we explain the variable-separation method for the composite dielectrics discussed in Chapters 2, 4, and 5 that have straight or linear interfaces and/or conductor surfaces. In this chapter, angles are given in radians unless otherwise stated.

T. Takuma and B. Techaumnat, *Electric Fields in Composite Dielectrics and their Applications*, Power Systems, DOI 10.1007/978-90-481-9392-9_8,

8.1.1 Dielectric Interface in Contact with a Plane Conductor

We consider the configuration of two dielectrics with dielectric constants (relative permittivities) ε_A and ε_B, the straight interface (boundary) of which contacts a plane conductor (electrode), as shown in Fig. 8.1. When we consider the field behavior in the close neighborhood of the point of contact, the problem can be treated as two-dimensional (2D). Thus we can express the electric potential in 2D polar coordinates (r, θ), as shown in Fig. 8.2, where the origin is taken as the point of contact on the grounded conductor. When the conductor is stressed at a fixed potential, the treatment is essentially the same: we simply add the constant value to the potentials below.

By applying the variable-separation method, potentials ϕ_A and ϕ_B in the two dielectric media are expressed with an infinite series of r^n and trigonometric functions that satisfies Laplace's equation. It is to be noted that the exponent n must not be negative to ensure a finite value of the potential near the origin (contact point), and that the smallest positive value of n is predominant near the point. Thus, if we use such a value of n, the potentials are expressed as follows:

$$\phi_A \cong r^n[a\cos(n\theta) + b\sin(n\theta)],$$

and

$$\phi_B \cong r^n[c\cos(n\theta) + d\sin(n\theta)] \tag{8.1}$$

where $a, b, c,$ and d are constants to be determined from the boundary conditions. As explained above, the constant component is omitted in these potentials, and it does not contribute essentially to the field behavior (singularity) near the contact point.

On the conductor surface, i.e., on the boundary $\theta = 0$, $\phi_A = 0$, and on the boundary $\theta = \pi$, $\phi_B = 0$, which leads to

$$\phi_A \cong ar^n\sin(n\theta) \quad \text{and} \quad \phi_B \cong br^n\sin[n(\pi - \theta)]. \tag{8.2}$$

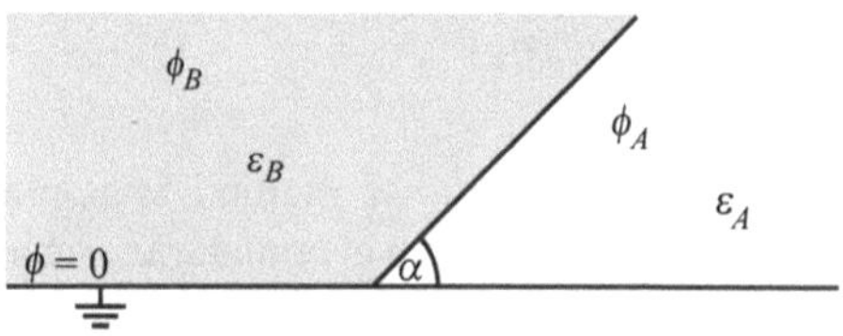

Fig. 8.1 Configuration of a straight dielectric interface in contact with a plane conductor (electrode) surface

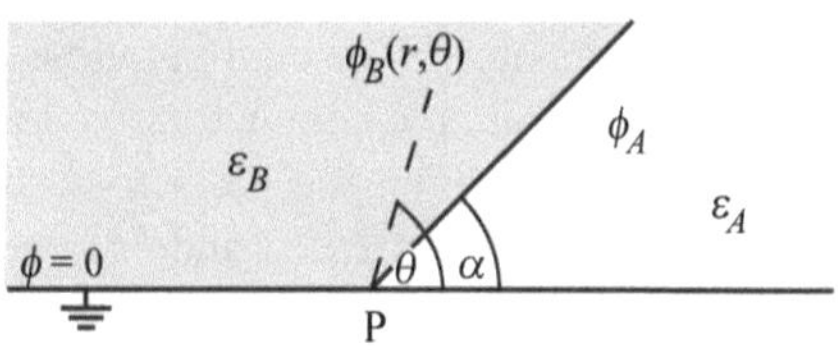

Fig. 8.2 Electric potentials in the configuration of Fig. 8.1 in 2D polar coordinates

Because the two components of the electric field are given as

$$E_r = -\frac{\partial \phi}{\partial r}$$

and

$$E_\theta = -\frac{1}{r}\frac{\partial \phi}{\partial \theta},$$

these components, and thus also the total field strength, are proportional to r^{n-1}.

At $\theta = \alpha$ (the dielectric interface), we impose the boundary conditions of the continuity of the potential as well as of the normal component of electric flux density:

$$\phi_A = \phi_B \quad \text{and} \quad \varepsilon_A \frac{\partial \phi_A}{\partial \theta} = \varepsilon_B \frac{\partial \phi_B}{\partial \theta}. \tag{8.3}$$

Substituting Eq. 8.2 into Eq. 8.3, we have the following transcendental equation for determining real-number exponent n [1]:

$$\varepsilon_B \tan(n\alpha) + \varepsilon_A \tan[n(\pi - \alpha)] = 0 \tag{8.4}$$

Values of n calculated from Eq. 8.4 are plotted in relation to α for several values of $\varepsilon_B/\varepsilon_A$ as a parameter in Fig. 2.3.

8.1.2 Solution of Exponent *n*

We consider the solutions of Eq. 8.4, which is the same as Eq. 2.3, based on the behavior of the following function, $f(n)$

$$f(n) = \frac{\varepsilon_B}{\varepsilon_A} \tan(n\alpha) + \tan[n(\pi - \alpha)] \tag{8.5}$$

Figure 8.3 represents the first and second terms of $f(n)$ in relation to n ranging from 0 to 4. For α lying between 0 and $\pi/2$, the first term changes from a positive infinity to a negative infinity at $n_1 = \pi/2\alpha$, where n_1 is larger than 1. On the other hand, the second term changes from a positive infinity to a negative infinity at $n_2 = \pi/2(\pi-\alpha)$, but n_2 is smaller than 1. As a result, $f(n)$ changes from a positive infinity to a negative infinity at $n_2 = \pi/2(\pi-\alpha)$, and increases monotonously in the range of n from n_2 to n_1. Because $f(1) = (\varepsilon_B/\varepsilon_A - 1) \cdot \tan\alpha$ and $\varepsilon_B/\varepsilon_A$ is larger than 1, $f(n)$ is positive at $n = 1$. It follows from these behaviors that $f(n)$ has the smallest solution between n_2 and 1.

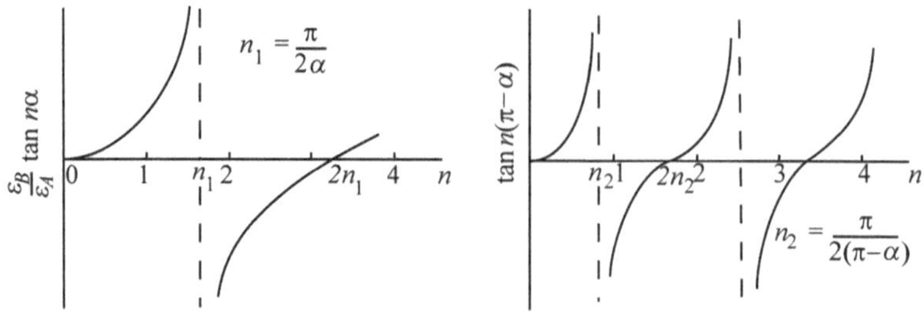

Fig. 8.3 The two terms of Eq. 8.5 in relation to n [2]

If $\varepsilon_B/\varepsilon_A$ is smaller than 1, $f(n)$ is negative at $n = 1$, which means that the solution of n below 1 vanishes and the smallest solution becomes larger than 1. In this case, the field strength approaches zero as the contact point is approached. A similar situation occurs when $\varepsilon_B/\varepsilon_A > 1$ and $\pi/2 < \alpha < \pi$. One of these cases is realized at point Q of Fig. 2.6.

8.1.3 *Two Dielectrics Without a Contacting Conductor*

When two dielectrics ε_A and ε_B make a straight interface without a contacting conductor, as shown in Fig. 8.4, the treatment is somewhat different from that described above in Section 8.1.1. By applying the variable-separation method, potentials ϕ_A and ϕ_B in the two dielectrics are expressed in the same manner as in Eq. 8.1. However, as no equipotential surface exists in this case, the boundary conditions are imposed at the two dielectric interfaces $\theta = 0$ for ϕ_A ($\theta = 2\pi$ for ϕ_B) and $\theta = 2\alpha$, in an approach similar to that of Eq. 8.3. This leads to simultaneous linear equations for constants a, b, c, and d as unknowns in Eq. 8.1. For these unknowns to have nonzero solutions, the following determinant of the (4×4) matrix composed of the corresponding coefficients for a, b, c, and d must be zero, i.e.,

$$\begin{vmatrix} 1 & 0 & -\cos B & -\sin B \\ \cos A & \sin A & -\cos A & -\sin A \\ 0 & \varepsilon_A & \varepsilon_B \sin B & -\varepsilon_B \cos B \\ -\varepsilon_A \sin A & \varepsilon_A \cos A & \varepsilon_B \sin A & -\varepsilon_B \cos A \end{vmatrix} = 0 \tag{8.6}$$

In this equation, $A = 2n\alpha$ and $B = 2\pi n$. The angle 2α is shown in Fig. 8.4. By arranging the terms here, we obtain the following expression for determining exponent n:

$$\left(\varepsilon_A^2 + \varepsilon_B^2\right) \sin A \sin(A - B) - 2\varepsilon_A 2\varepsilon_B \{1 - \cos A \cos(A - B)\} = 0 \tag{8.7}$$

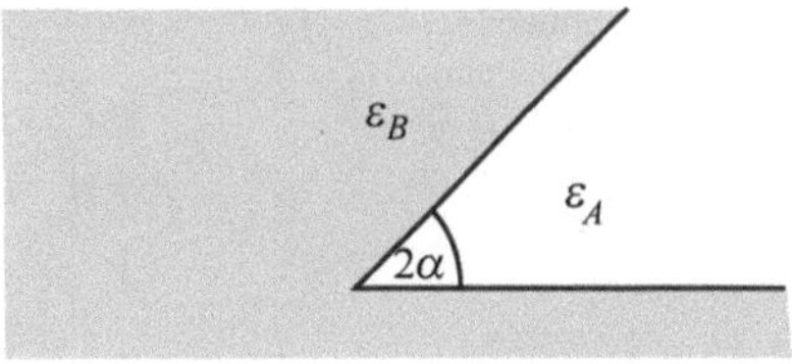

Fig. 8.4 Two dielectrics with straight interfaces without a contacting conductor

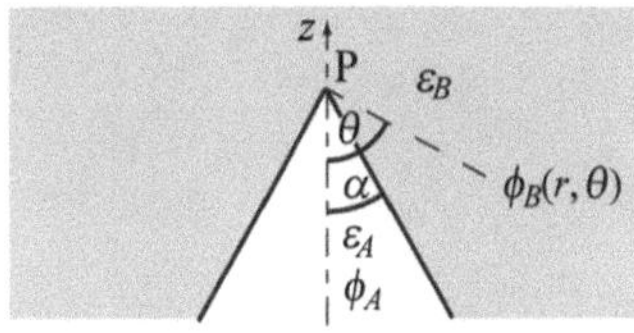

Fig. 8.5 Dielectric projection or void in an axisymmetric configuration

This expression can be further transformed to the following simpler one:

$$4(\varepsilon_A \cos C \sin D - \varepsilon_B \sin C \cos D)(\varepsilon_B \cos C \sin D - \varepsilon_A \sin C \cos D) = 0,$$

where $C = A/2 = n\alpha$ and $D = (A{-}B)/2 = n(\alpha{-}\pi)$.
As a result, we obtain the final equation [3]:

$$\frac{\tan D}{\tan C} = \frac{\varepsilon_B}{\varepsilon_A}, \text{ or } \frac{\varepsilon_A}{\varepsilon_B}, \tag{8.8}$$

which corresponds to Eq. 8.4, as explained above in Section 8.1.1.

8.1.4 Axisymmetric Case

We consider an axisymmetric (AS) configuration, as shown in Fig. 8.5, where a cone-shaped protrusion (or projection) or void (if medium ε_A is a gas or a vacuum) exists under otherwise uniform field E_0. E_0 is in the direction of the z-axis, thus making the field axially symmetrical in respect to the z-axis. In a similar approach to those in the 2D cases explained in Section 8.1.1, potentials ϕ_A and ϕ_B in the two dielectrics are expressed using an infinite series of r^n multiplied by Legendre functions that satisfies Laplace's equation in the AS polar coordinates. The smallest positive value of n is predominant in the close vicinity of apex P. If we use such a value of n, the potentials are expressed as follows:

$$\phi_A \cong a r^n P_n(\cos\theta),$$

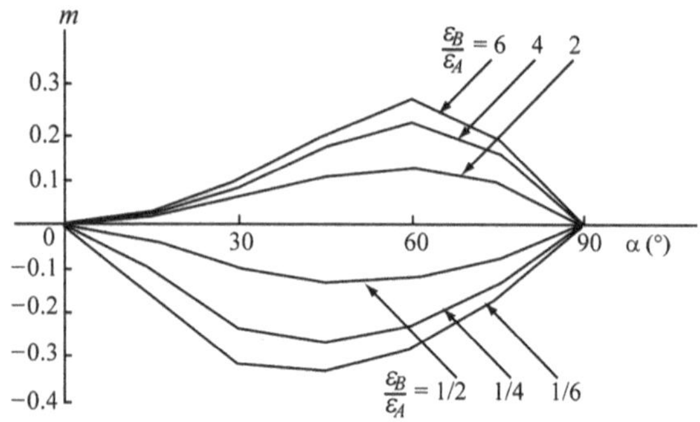

Fig. 8.6 Value of m (= n – 1) versus contact angle α [1] . © 1978 IEEE

and

$$\phi_B \cong br^n P_n(\cos(\pi - \theta)) = br_n P_n(-\cos\theta), \tag{8.9}$$

where P_n denotes an nth order Legendre function. Just as in the 2D cases, the two components of the field strength and the resultant total value are all proportional to r^{n-1}. From the boundary condition at $\theta = \alpha$, we can derive the following equation for determining n [1]:

$$\begin{aligned} &\varepsilon_B P_n(\cos\alpha)[\cos\alpha P_n(-\cos\alpha) + P_{n-1}(-\cos\alpha)] \\ &\quad - \varepsilon_A P_n(-\cos\alpha)[\cos\alpha P_n(\cos\alpha) - P_{n-1}(\cos\alpha)] = 0 \end{aligned} \tag{8.10}$$

Figure 8.6 shows m (= n – 1) at 15° intervals of α for several values of $\varepsilon_B/\varepsilon_A$, as obtained by solving Eq. 8.10. The graphs show lines which connect the discretely computed values. As expected, m is positive or negative according to whether $\varepsilon_B/\varepsilon_A$ is larger or smaller than unity. This corresponds to the self-explanatory fact that the field strength at P is infinitely high for $\varepsilon_B/\varepsilon_A < 1$ and zero for $\varepsilon_B/\varepsilon_A > 1$. We can also determine the values of m directly from the slopes of plots of the numerically calculated electric field versus distance in log-log scales. Values based on the results obtained by the charge simulation method (CSM) agreed very well with the analytical values from Eq. 8.10 shown in Fig. 8.6 [1].

8.1.5 Configurations with Three Dielectrics

We consider three dielectrics ε_A, ε_B, and ε_C meeting at a common point with straight interfaces, as shown in Fig. 8.7. The treatment is similar to the configurations of two dielectrics explained in Section 8.1.3. By applying the variable-separation method, potentials ϕ_A, ϕ_B, and ϕ_C in the three media are expressed as an infinite series of r^n and trigonometric functions. Also in this case, exponent n must not be negative to

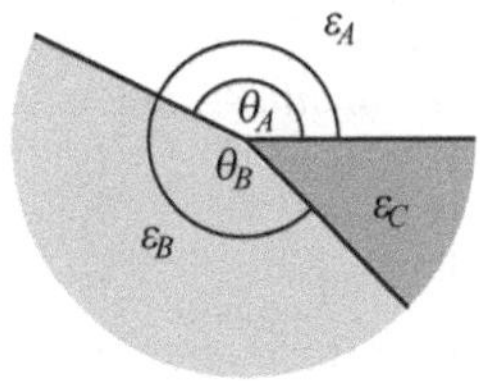

Fig. 8.7 Three dielectrics with straight interfaces contacting at a point (an edge)

ensure a finite value of the potential near the origin (contact point or edge), and the smallest positive value of n is predominant near the edge. Thus, if we use such a value of n:

$$\phi_A \cong r^n (a \cos n\theta + b \sin n\theta),$$

$$\phi_B \cong r^n (c \cos n\theta + d \sin n\theta),$$

and

$$\phi_C \cong r^n (e \cos n\theta + f \sin n\theta), \tag{8.11}$$

where a, b, c, d, e, and f are constants to be determined from the boundary conditions. As explained in Section 8.1.1, the constant component of these potentials is omitted here, and it does not essentially contribute to the field behavior or singularity near the point of contact.

On the three interfaces $\theta = \theta_A$, $\theta = \theta_B$, and $\theta = 0$ (2π for ϕ_C), we impose the boundary conditions of the continuity of the potential as well as of the normal component of the electric flux density, as follows:
At $\theta = \theta_A$,

$$\phi_A = \phi_B \text{ and } \varepsilon_A \frac{\partial \phi_A}{\partial \theta} = \varepsilon_B \frac{\partial \phi_B}{\partial \theta},$$

at $\theta = \theta_B$,

$$\phi_B = \phi_C \text{ and } \varepsilon_B \frac{\partial \phi_B}{\partial \theta} = \varepsilon_C \frac{\partial \phi_C}{\partial \theta},$$

at $\theta = 0$ (2π for ϕ_C)

$$\phi_C = \phi_A \text{ and } \varepsilon_C \frac{\partial \phi_C}{\partial \theta} = \varepsilon_A \frac{\partial \phi_A}{\partial \theta}. \tag{8.12}$$

These conditions lead to six simultaneous linear equations for a, b, c, d, e, and f as unknowns. In order that these unknowns have nonzero solutions, the following determinant of the (6 × 6) matrix composed of the coefficients for a, b, c, d, e, and f must be zero [4]:

$$\begin{vmatrix} \cos n\theta_A & \sin n\theta_A & -\cos n\theta_A & -\sin n\theta_A & 0 & 0 \\ \varepsilon_A \sin n\theta_A & -\varepsilon_A \cos n\theta_A & -\varepsilon_B \sin n\theta_A & \varepsilon_B \cos n\theta_A & 0 & 0 \\ 0 & 0 & \cos n\theta_B & \sin n\theta_B & -\cos n\theta_B & -\sin n\theta_B \\ 0 & 0 & \varepsilon_B \sin n\theta_B & -\varepsilon_B \cos n\theta_B & -\varepsilon_C \sin n\theta_B & \varepsilon_C \cos n\theta_B \\ -1 & 0 & 0 & 0 & \cos 2\pi n & \sin 2\pi n \\ 0 & \varepsilon_A & 0 & 0 & \varepsilon_C \sin 2\pi n & -\varepsilon_C \cos 2\pi n \end{vmatrix} = 0. \tag{8.13}$$

It should be noted that in the case of two dielectrics explained in Section 8.1.3, the same procedure leads to the (4×4) matrix of Eq. 8.6. By reforming and rearranging Eq. 8.13, we can derive the following expression:

$$\begin{aligned} &\left(\varepsilon_A{}^2 + \varepsilon_B{}^2\right) \varepsilon_C \sin A \sin(A - B) \cos(B - C) \\ &\quad + \left(\varepsilon_A{}^2 + \varepsilon_C{}^2\right) \varepsilon_B \sin A \cos(A - B) \sin(B - C) \\ &\quad - \left(\varepsilon_B{}^2 + \varepsilon_C{}^2\right) \varepsilon_A \cos A \sin(A - B) \sin(B - C) \\ &\quad - 2\varepsilon_A \varepsilon_B \varepsilon_C \{1 - \cos A \cos(A - B) \cos(B - C)\} = 0 \end{aligned} \tag{8.14}$$

In this equation, $A = n\theta_A$, $B = n\theta_B$, and $C = 2\pi n$.

Applying the Mathematica, on the other hand, gives the following expression, a linear combination of cosine functions of the angles:

$$\begin{aligned} 8\varepsilon_A \varepsilon_B \varepsilon_C &- (\varepsilon_A + \varepsilon_B)(\varepsilon_A + \varepsilon_C)(\varepsilon_B + \varepsilon_C) \cos C \\ &+ (\varepsilon_A - \varepsilon_B)(\varepsilon_A - \varepsilon_C)(\varepsilon_B + \varepsilon_C) \cos(2A - C) \\ &+ (\varepsilon_A + \varepsilon_B)(\varepsilon_A - \varepsilon_C)(\varepsilon_B - \varepsilon_C) \cos(2B - C) \\ &- (\varepsilon_A - \varepsilon_B)(\varepsilon_A + \varepsilon_C)(\varepsilon_B - \varepsilon_C) \cos(2A - 2B + C) = 0 \end{aligned} \tag{8.15}$$

Although Eq. 8.15 appears quite different from Eq. 8.14, which consists of the cubic multiplication of trigonometric functions, these two equations are equivalent. Values of n cannot be given explicitly except in some special cases, but we can apply an appropriate numerical method, such as the Newton–Raphson method, to solve these equations.

Section 4.3 compares the values of n obtained from Eq. 8.14 or 8.15 with the results from numerical field calculations using the CSM. A comparison has been made only for the limited case of Fig. 4.4 with the dielectric constants $\varepsilon_A = 1.0$, $\varepsilon_B = 2.3$, and $\varepsilon_C = 6.0$, but Fig. 4.6 shows an excellent coincidence between the analytical and numerically computed values for a wide range of contact angles.

8.2 Iterative Image Charge Method

In the simple image charge method, a single charge or a finite number of discrete charges are used in place of the real charge distribution to equivalently express the electric field in a part of the whole domain. Some typical examples are a point or line charge situated at a point (mirror image point) perpendicularly below and at the same distance from a conducting plane, a point charge for a conducting sphere, and a line charge for a conducting cylinder. These are explained in any textbook on electromagnetism. Here we explain the image charge method for the composite dielectrics considered in Chapter 3, in which an infinite number of image charges are iteratively arranged.

8.2.1 Conducting Sphere on a Solid Dielectric Plane

The first example is an axisymmetric (AS) configuration as shown in Fig. 8.8, namely, a conducting sphere in contact with a solid dielectric plane in another (dielectric) medium. The surrounding medium is usually a gas or a vacuum, and its dielectric constant (relative permittivity) is denoted by ε_A. The dielectric constant of the solid is ε_B. It is to be noted that we cannot easily apply an image charge method when a solid dielectric has finite thickness.

We consider a conducting sphere with radius R at potential V, as shown in Fig. 8.8. We can represent the field with an infinite number of image charges repetitively located inside the sphere and the dielectric ε_B, as follows. We take the point of contact as the origin of the AS (r–z) coordinates. First, point charge Q is situated at the sphere center ($z = R$) to satisfy the boundary condition on the sphere surface, where $Q = 4\pi\varepsilon_A\varepsilon_0 RV$. If we define variable P as in Eq. 8.16, then an image charge of magnitude $-QP$ situated at corresponding image point $z = -R$ inside the dielectric ε_B satisfies the boundary condition on the plane interface.

$$P = (\varepsilon_s - 1)/(\varepsilon_s + 1), \varepsilon_s = \varepsilon_B/\varepsilon_A. \tag{8.16}$$

Table 8.1 gives a series of positions (z-coordinates) and magnitudes of these image charges thus successively located inside the conducting sphere and the solid dielectric ε_B. By summing the field resulting from these charges, the field strength E_m (maximal value) at the contact point on the ε_A side can be given as

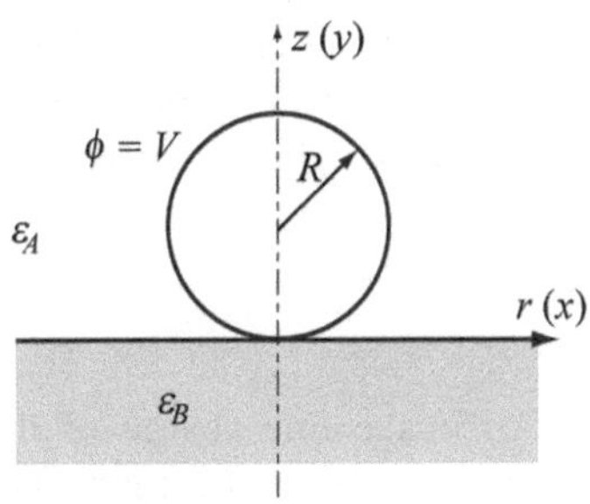

Fig. 8.8 Spherical (or cylindrical) conductor lying on a solid dielectric plane of infinite thickness

Table 8.1 Position (z-coordinate) and magnitude of successive image charges located on the z-axis for the sphere-to-solid dielectric plane (AS) configuration of Fig. 8.8 [5, 6]. © 1984 IEEE

In the sphere	z	R	$R/2$	$R/3$	...	R/n
	Magnitude	Q	$PQ/2$	$P^2Q/3$	...	$P^{n-1}Q/n$
In the solid dielectric	z	$-R$	$-R/2$	$-R/3$	...	$-R/n$
	Magnitude	$-PQ$	$-P^2Q/2$	$-P^3Q/3$	...	$-P^nQ/n$

$$E_m = \frac{1}{4\pi\varepsilon_A\varepsilon_0}\left\{\sum_{n=1}^{\infty}\frac{QP^{n-1}}{n}\left(\frac{n}{R}\right)^2 + \sum_{n=1}^{\infty}\frac{QP^n}{n}\left(\frac{n}{R}\right)^2\right\} = \frac{V}{R}\left[\frac{\varepsilon_s(\varepsilon_s+1)}{2}\right]. \quad (8.17)$$

Although only an infinitesimal gap exists on the ε_A side at the contact point, the field is realizable on the ε_B side and its strength is equal to E_m/ε_s.

When $\varepsilon_s = 1$, E_m is V/R, as expected for a conducting sphere existing isolated in space. Therefore, we can express the field enhancement with the contact point strength E_1 ($= V/R$) as

$$E_m = \frac{\varepsilon_s(\varepsilon_s+1)}{2}E_1$$

on the ε_A side, and

$$E_m = \frac{\varepsilon_s+1}{2}E_1 \quad (8.18)$$

on the ε_B side.

8.2.2 Conducting Cylinder on a Solid Dielectric Plane

Similar calculations using an infinite number of image line charges give the field behavior for the two-dimensional (2D) case of a conducting cylinder at potential V lying on a solid dielectric plane, as shown in Fig. 8.8. In this case, however, we first consider the dielectric plate of finite thickness D on the grounded conductor plane, as shown in Fig. 3.8. We take the point of contact as the origin of the 2D (x–y) coordinates. For $\varepsilon_B = \varepsilon_A$ and when D is sufficiently large, a line charge of density λ and its image are situated, respectively, on the axis of the cylinder (at $y = R$) and below the conductor plane (at $y = -R-2D$) to satisfy the boundary condition at $y = 0$ on the cylinder surface. λ is related to the cylinder voltage V as

$$V = \frac{\lambda}{2\pi\varepsilon_A\varepsilon_0}\ln\frac{R+2D}{R}. \quad (8.19)$$

An image line charge of density $-P\lambda$ situated at $y = -R$ inside the solid dielectric satisfies the boundary condition on the solid surface, where P is given

Table 8.2 Position (y-coordinate) and magnitude (charge density) of successive image charges located on the y-axis for the cylinder-to-solid dielectric plane (2D) configuration of Fig. 8.8

In the cylinder	y	R	$R/2$	$R/3$	...	R/n
	Magnitude	λ	$P\lambda$	$P^2\lambda$	...	$P^{n-1}\lambda$
In the solid dielectric	y	$-R$	$-R/2$	$-R/3$	...	$-R/n$
	Magnitude	$-P\lambda$	$-P^2\lambda$	$-P^3\lambda$	...	$-P^n\lambda$

by Eq. 8.16. Table 8.2 gives a series of positions (y-coordinates) and densities for these image charges in the 2D configuration corresponding to those in Table 8.1 for the AS case. Thus, when D is sufficiently large, the field strength E_m (maximal value) at the contact point on the ε_A side is

$$E_m = \frac{\lambda}{2\pi\varepsilon_A\varepsilon_0}\sum\left(P^{n-1}\lambda\frac{n}{R} + P^n\lambda\frac{n}{R}\right) = \frac{\lambda}{2\pi\varepsilon_A\varepsilon_0}\frac{1}{R}\frac{\varepsilon_s(\varepsilon_s+1)}{2} \tag{8.20}$$

Since the field strength is $\lambda/(2\pi\varepsilon_A\varepsilon_0 R)$ on the cylinder surface when $\varepsilon_s = 1$ and D is sufficiently large, the relationship of Eq. 8.18 also holds for the 2D cylinder-to-dielectric plane configuration.

8.2.3 *Conducting Sphere or Cylinder Separated from a Dielectric Plane*

We here explain the image charge method for the configuration of a conducting sphere or cylinder not in contact with the solid dielectric (ε_B) plane but separated by a gap from it. The separation distance is S, as shown in Fig. 8.9. The basic procedure of locating image charges alternately inside the conductor and inside the dielectric ε_B is the same as explained in Sections 8.2.1 and 8.2.2 for cases without a gap.

For an AS configuration shown in Fig. 8.9, the position (z-coordinate z_n) and magnitude Q_n of the nth image charge are given as follows [7]:

Inside the sphere: $z_n = R + S - \dfrac{R^2}{(R + S + z_{n-1})}$, where $z_1 = R + S$

$$Q_n = QP^{n-1}\sinh\alpha/\sinh(n\alpha) \tag{8.21}$$

Inside dielectric ε_B: $z'_n = -z_n$

$$Q'_n = -QP^n\sinh\alpha/\sinh(n\alpha) \tag{8.22}$$

In these equations, $Q = 4\pi\varepsilon_A\varepsilon_0 RV$ and P is given by Eq. 8.16, while

$$\cosh\alpha = 1 + S/R \tag{8.23}$$

Thus, adding the fields induced by these charges leads to the field strength at any point. For example, E_P at point P in Fig. 8.9 on the dielectric plane is given by

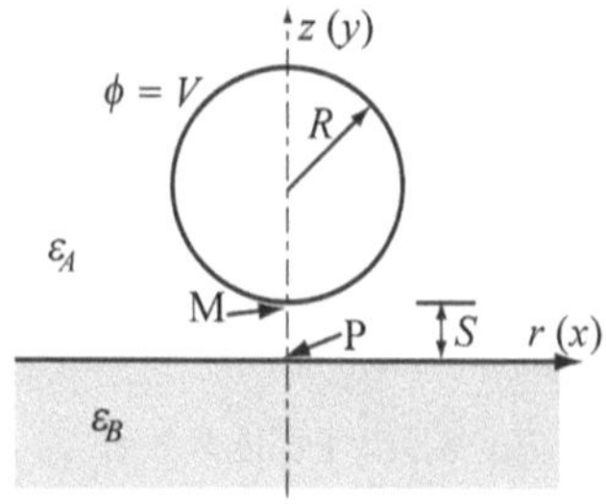

Fig. 8.9 Spherical (cylindrical) conductor lying at distance S from a solid dielectric plane

$$E_P = \frac{Q}{4\pi\varepsilon_A\varepsilon_0}(1+P)\left[\frac{1}{(R+S)^2} + \sum_{n=2}^{\infty}\frac{P^{n-1}}{z_n^2}\frac{\sinh\alpha}{\sinh(n\alpha)}\right] \tag{8.24}$$

The image charge method can be similarly applied for the 2D case of a conducting cylinder lying separated from the dielectric plane with gap S, as shown in Fig. 8.9. An infinite series of line charges of density λ alternately located inside the cylinder and inside the dielectric gives the following field strength at P:

$$E_P = \frac{\lambda}{2\pi\varepsilon_A\varepsilon_0}(1+P)\left[\frac{1}{R+S} + \sum_{n=2}^{\infty}\frac{P^{n-1}}{z_n^2}\right] \tag{8.25}$$

Some typical results computed for field strength E_M (the value at M on the conductor tip) relative to S are explained in Section 3.1.3.

8.3 Uncharged Conducting Sphere Under a Uniform Field on a Dielectric Plane

8.3.1 Image Charges

We consider the configuration of an uncharged conducting sphere of radius R on a dielectric interface (dielectric slab of infinite thickness), as shown in Fig. 8.10. The potential of the conductor is not given but varies with externally applied field E_0. The dielectric constant (relative permittivity) is ε_A for the (background) space surrounding the conductor and ε_B for the dielectric slab. The image charge method for this configuration is more complicated than the method described in Section 8.2, and therefore it is explained separately in this section. More details are given by B. Techaumnat and T. Takuma [8].

The first scheme is to locate an image charge for a configuration consisting of a grounded conducting sphere and point charge q or dipole charge p as shown in Fig. 8.11 which satisfies the zero-potential condition of the conductor in medium

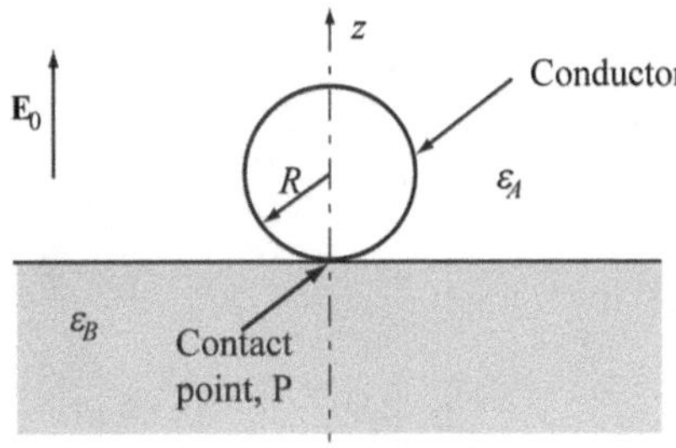

Fig. 8.10 Uncharged conducting sphere under a uniform field

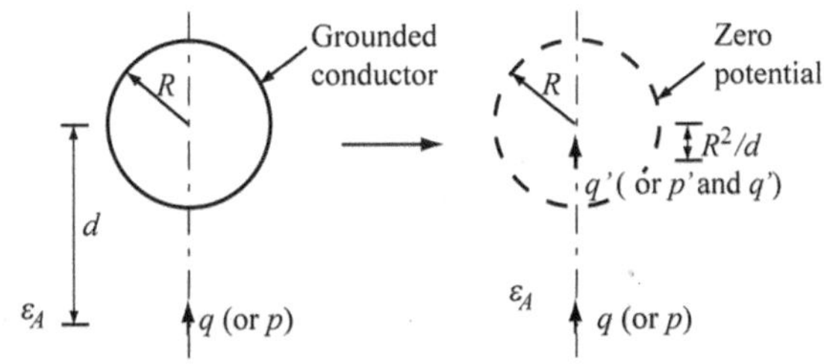

Fig. 8.11 Point charge q or dipole charge p and a grounded conducting sphere

ε_A. The former case of a point charge is analyzed for the configuration of Fig. 8.8, and the image scheme is well known. In the case of dipole charge p, two images, point charge q' and dipole charge p', must be incorporated to satisfy the zero-potential condition. As shown in Fig. 8.11, they are located at the same place as the point charge, and their magnitudes are

$$q' = -\frac{R}{d^2}p, \quad \text{and} \quad p' = \left(\frac{R}{d}\right)^3 p. \tag{8.26}$$

When the conducting sphere is not at zero potential but is electrically floating, as shown in Fig. 8.10, we add point charge $-q'$ at the sphere center to negate the net charge of the conductor.

The second image scheme is shown in Fig. 8.12, where point charge q or dipole charge p is located at height d above the planar interface of two dielectrics ε_A and ε_B. The scheme for a point charge is well known and the positions of the images q' and q'' are given in Fig. 8.12b. For the case of dipole charge p, its images p' and p'' are at the same positions as q' and q'', respectively. The image magnitudes are related to the sources as follows:

$$p' = Pp, \quad \text{and} \quad p'' = (1 - P)p, \tag{8.27}$$

where P is given by Eq. 8.16.

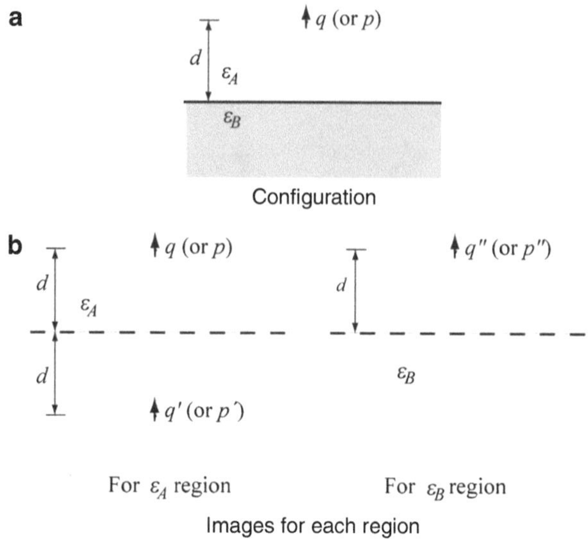

Fig. 8.12 Point or dipole charge and a planar dielectric interface

8.3.2 *Procedure of Image Charge Location*

The potential in the configuration of Fig. 8.10 can be obtained by using the following procedure.

8.3.2.1 First Set of Images

The solution begins with an image induced by the externally applied field E_0, which is dipole charge

$$p^{(1)} = 4\pi\varepsilon_A\varepsilon_0 E_0 R^3 \tag{8.28}$$

at the center of the conducting sphere. This dipole results in image charges as described by Eqs. 8.26 and 8.27, which further induce images with respect to the conductor, and so on. Images for the potential in the ε_A region are listed in Table 8.3, where $q^{(i)}$ denotes an image point charge, and some of their positions are shown in Fig. 8.13. The superscript (i) denotes the step of repetition. At the contact point between the conductor and the dielectric slab, the electric field E_{c1} due to all p and p' can be written as

$$E_{c1} = \sum_{i=1}^{\infty} \left\{ \frac{2p^{(i)}}{\left[h^{(i)}\right]^3} + \frac{2p'^{(i)}}{\left[h^{(i)}\right]^3} \right\} + 2E_0 \frac{1+P}{1-P} \tag{8.29}$$

Table 8.3 First set of image charges induced from dipole $p^{(1)}$ at the sphere center [8].

(i)	$h^{(i)}/R$	$q^{(i)}/p^{(1)}$	$p^{(i)}/p^{(1)}$	$p'^{(i)}/p^{(1)}$	$d^{(i)}/R$
1	1	–	1	P	2
2	1/2	$-(1/2)^2P/R$	$(1/2)^3P$	$(1/2)^3P^2$	3/2
3	1/3	$-(1/2)(1/3)^2P^2/R$	$(1/3)^3P^2$	$(1/3)^3P^3$	4/3
⋮					
i	$1/i$	$-[1/(i-1)](1/i)^2P^{i-1}/R$	$(1/i)^3P^{i-1}$	$(1/i)3P^i$	$(i+1)/i$
⋮					

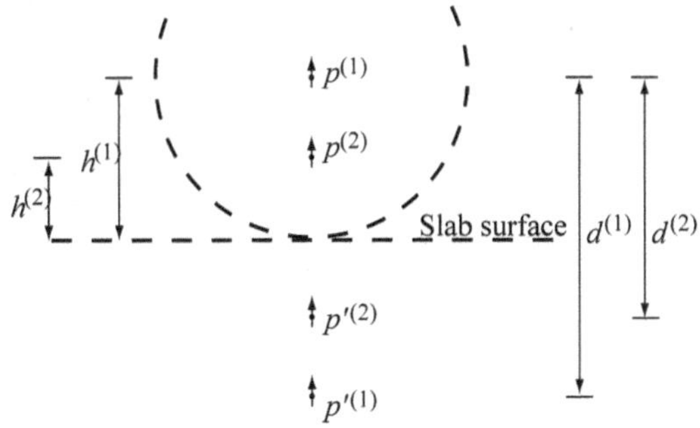

Fig. 8.13 Positions of the first image set

8.3.2.2 Second Set of Images

The images originate from the point charge $q^{(i)}$ ($i = 2, 3, \ldots$) from the first step. Let point charge q_m at height R/m ($m \geq 2$) represent $q^{(i)}$, then q_m results in another sequence of images, as given in Table 8.4, where the superscript (j) denotes the step of repetition. Examples of the positions of these images for $j = 1$ and 2 are shown in Fig. 8.14. The electric field E_{c2} at the contact point resulting from all q_m ($2 \leq m < \infty$) and their images is given by

$$E_{c2} = E_0 \frac{1+P}{(1-P)^2} P. \tag{8.30}$$

8.3.2.3 Third Set of Images

The last set of images fulfills the zero net charge condition of the conducting sphere. It begins with point-charge image q'' at the sphere center which negates all the point charges shown in Table 8.4, i.e.,

$$q'' = -\sum_{m=2}^{\infty}\sum_{j=1}^{\infty} q_m^{(j)} = \sum_{m=2}^{\infty} q''_m, \tag{8.31}$$

Table 8.4 Second set of images resulting from point charge q_m at a height of R/m [8] © 2005 Elsevier B.V.

(j)	$h^{(j)}/R$	$q_m^{(j)}/q_m$	$q'_m{}^{(j)}/q_m$	$d^{(i)}/R$
1	$1/m$	1	$-P$	$(m+1)/m$
2	$1/(m+1)$	$[m/(m+1)]P$	$-[m/(m+1)]P^2$	$(m+2)/(m+1)$
3	$1/(m+2)$	$[m/(m+2)]P^2$	$-[m/(m+2)]P^3$	$(m+3)/(m+2)$
⋮				
j	$1/(m+j-1)$	$[m/(m+j-1)]P^{j-1}$	$-[m/(m+j-1)]P^j$	$(m+j)/(m+j-1)$
⋮				

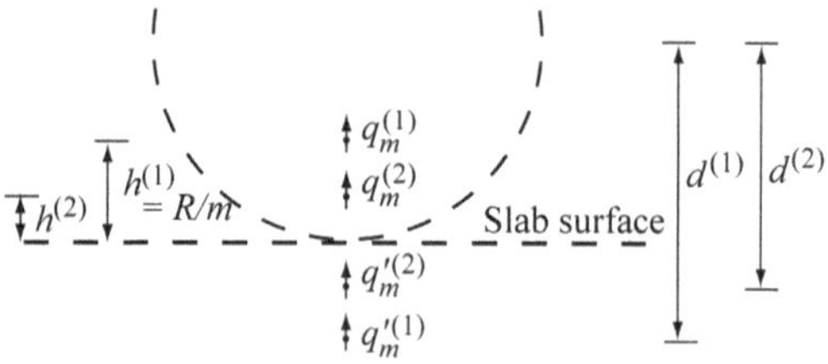

Fig. 8.14 Positions of the second image set

where q''_m negates all $q_m^{(i)}$, of which the relationship to q_m is shown in Table 8.4. Replacing q_m by $q^{(i)}$ in Table 8.3 with the index $i = m$, we obtain

$$q''_m = -\sum_{j=1}^{\infty} \frac{m}{m+j-1} P^{j-1} q_m = \frac{p^{(1)}}{R} \frac{1}{P} \left[\frac{1}{m(m-1)}\right] \left[\ln \frac{1}{1-P} - \sum_{j=1}^{m-1} \frac{P^j}{j}\right]. \quad (8.32)$$

From Eqs. 8.31 and 8.32,

$$q'' = \frac{p^{(1)}}{R} \frac{1}{P} \left[\ln\left(\frac{1}{1-P}\right) - \beta_P\right], \quad (8.33)$$

where $\beta_P = \sum_{j=1}^{\infty} P^j / j^2$.

The image charge q'' leads to another sequence of images, as shown in Table 8.5. From the table, it can be deduced that the sum of the point charges at $h = R/i$ is equal to $\alpha_P P^{i-1} q''/i$, where

$$\alpha_P = \sum_{k=1}^{\infty} S_k, \quad (8.34)$$

Table 8.5 Third set of images originating from unit point charge q'' at the sphere center [8].

(k)	Images inside the sphere				Images inside the dielectric slab			
	$h = R$	$R/2$	$R/3$	$R/4$	$h = -R$	$-R/2$	$-R/3$	$-R/4$
1	1				$-P$			
2	$-P/2$	$P/2$			$P^2/2$	$-P^2/2$		
3	$-P^2/12$	$-P^2/4$	$P^2/3$		$P^3/12$	$P^3/4$	$-P^3/3$	
4	$-P^3/24$	$-P^3/24$	$-P^3/6$	$P^3/4$	$P^4/24$	$P^4/24$	$P^4/6$	$-P^4/4$
⋮								

$$S_k = -\sum_{l=0}^{k-1} \frac{P^{k-l}}{k-l+1} S_l, \quad \text{and} \quad S_0 = 1. \tag{8.35}$$

At the contact point, the electric field due to q'' in Eq. 8.33 and all of its images is

$$E_{c3} = -E_0 \frac{\alpha_P}{P} \frac{1+P}{(1-P)^2} \left[\ln\left(\frac{1}{1-Pp}\right) - \beta_P \right]. \tag{8.36}$$

Finally, the electric field at the contact point in the configuration of Fig. 8.10 can be determined from the sum of E_{c1}, E_{c2}, and E_{c3}.

8.4 Re-expansion Method for a System of Particles

8.4.1 Principle

We regard the re-expansion method as an analytical method because the field is explicitly given as an infinite series of spherical harmonics. This method is applicable to potential problems in configurations consisting of spherical and planar objects. Although the use of a computer is inevitable to perform a complicated iterative computation, the method has advantages such as shorter calculation time and more accurate results compared to purely numerical methods.

The re-expansion method described in this section solves potential problems in an iterative manner based on image schemes [9, 10]. On the other hand, the so-called fast multipole method (FMM) proposed by V. Rohklin [11] is also based on the three re-expansion (translation) formulae, i.e., multipole to multipole expansion (M2M), multipole to local expansion (M2L) and local to local expansion (L2L), without image schemes. (For the latter method, see, for example, S. Hamada et al. [12], M. Kitano et al. [13], and T. Takuma and S. Hamada [14].)

As mentioned above, the re-expansion method utilizes an infinite series of spherical harmonics satisfying Laplace's equation to express electric potentials. The calculation principle is to find appropriate coefficients that fulfill all the boundary conditions by the sum of the harmonics. In expressing electric potential for a sphere centered at point **c**, we shall take **c** as the origin and denote spherical coordinates of a point related to **c** by (r_c, θ_c, φ_c). Similar notations will be used consistently hereafter for indicating the spherical coordinates of a point and the origin. The electric potential ϕ at point **p** (r_c, θ_c, φ_c) inside the sphere is then expanded about **c** as

$$\phi = \sum_{n=0}^{\infty} \sum_{m=-n}^{n} L_{n,m} r_c^n \bar{P}_{n,|m|}(\cos\theta_c)\exp(jm\varphi_c). \tag{8.37}$$

In this equation, $L_{n,m}$ is the potential coefficient, $\bar{P}_{n,|m|}$ the Legendre function normalized by $\sqrt{(n+|m|)/(n-|m|)!}$, and $j = \sqrt{-1}$. Equation 8.37 is the local expansion about the sphere center of the potential due to all sources outside the sphere, including free or polarization charge on its surface. For a simple case in which the sphere is a conductor, the potential expression reduces to $\phi = L_{0,0}$ (= constant). An important point to note is that we assume no charge inside the sphere.

On the other hand, if **p** is external to the sphere, the potential takes a different form:

$$\phi = \sum_{n=0}^{\infty} \sum_{m=-n}^{n} \left(\frac{A_{n,m}}{r_c^{n+1}} + M_{n,m} r_c^n \right) \bar{P}_{n,|m|}(\cos\theta_c)\exp(jm\varphi_c) \tag{8.38}$$

where $A_{n,m}$ and $M_{n,m}$ are potential coefficients. The terms with $A_{n,m}$ represent the potential due to charge on the sphere surface, which vanishes at infinite r_c. Terms with $M_{n,m}$ correspond to charge at other locations. For the local expansion (of $M_{n,m}$) in Eq. 8.38 to converge, the charge must be farther than **p** from the sphere center. We may regard $A_{n,m}$ in Eq. 8.38 as a multipole of orders (n, m) at the sphere center. The potential due to a unit multipole of orders (n, m) located at the origin is defined by

$$\phi = \frac{1}{r^{n+1}} \bar{P}_{n,|m|}(\cos\theta)\exp(jm\varphi). \tag{8.39}$$

Accordingly, charge on the sphere surface is re-expanded about **c** in the form of the multipole potential given in Eq. 8.38. Note that the definition of the unit multipole strength is slightly different from the conventional one in order to simplify the expression of the potential. A general formula for multipole re-expansion is described as follows.

For a unit multipole of orders (n, m) located at point **p**, as shown in Fig. 8.15, the potential is expressed as

$$\phi = \frac{1}{r_p^{n+1}} \bar{P}_{n,|m|}(\cos\theta_p)\exp(jm\varphi_p). \tag{8.40}$$

Fig. 8.15 A unit multipole at **p** and the spherical coordinates (D_q, α_q, β_q) relative to the origin point **q**

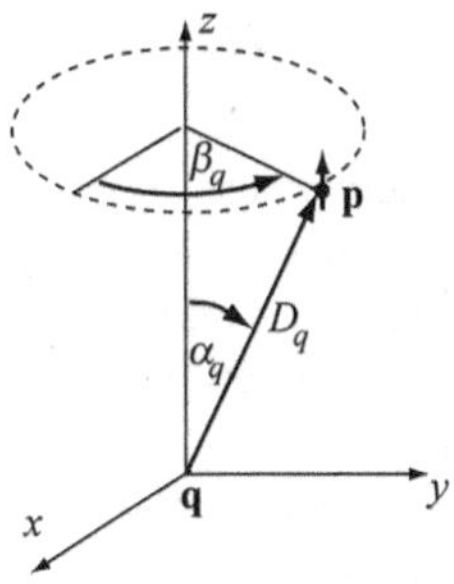

Re-expanding this potential locally about point **q** yields the following expression [15]

$$\phi = \sum_{j=0}^{\infty} \sum_{k=-j}^{j} M_{j,k} r_q^j \bar{P}_{j,|k|}(\cos\theta_q)\exp(\mathrm{j}k\varphi_q)$$

$$M_{j,k} = A_{n,m}^{j,k} \frac{(-1)^n}{D_q^{j+n+1}} \bar{P}_{n+j,|m-k|}(\cos\alpha_q)\exp\left[\mathrm{j}(m-k)\beta_q\right], \tag{8.41}$$

where (D_q, α_q, β_q) are the coordinates of **p** relative to **q**, and

$$A_{n,m}^{j,k} = \mathrm{j}^{|k-m|-|k|-|m|}\sqrt{\frac{(n+m+j-k)!(n-m+j+k)!}{(n+m)!(j-k)!(n-m)!(j+k)!}} \tag{8.42}$$

At point **p** on the surface of the sphere, the potential can be expressed by either Eq. 8.37 or 8.38. It follows that we may use boundary conditions of the potential and the normal electric field to relate the potential coefficients A and L to M for each (j, k) and construct a system of linear equations for solving the coefficients if M is given. An alternative is to determine the potential coefficients in an iterative manner by applying proper image schemes. The advantages of the iterative approach are (a) the small memory requirement and (b) the adjustable accuracy of solutions according to the number of iterations used.

It is to be noted here that the equations are described in this section with a general notation for 3D cases. For axisymmetric configurations, however, all the potential coefficients associated with nonzero k vanish. For example, Eq. 8.37 simply reduces to

$$\phi = \sum_{n=0}^{\infty} L_{n,0} r_c^n P_n(\cos\theta_c). \tag{8.43}$$

The normalized Legendre function becomes the simpler $P_n(\cos\,\theta_c)$ depending only on parameter n in Eq. 8.43, and the exponential terms disappear from the expression.

8.4.2 Image Schemes

Each image scheme described here is used to determine the coefficients of the potential that satisfy the boundary conditions on an object. If only one source multipole and one object are involved in the problem, then an appropriate image scheme can be applied only once to give the solutions. For a case in which more than one object exists, the schemes are applied repetitively to their corresponding interfaces until the solutions converge. In this case, the multipole re-expansion in Eq. 8.41 is usually employed to express a potential in the desired form, to which the image schemes can be applied. Some simple cases are explained below.

8.4.2.1 Grounded Plane

Consider multipole $B_{n,m}$ of orders (n, m) at point $\mathbf{p}$ above a planar interface separated by distance d, as shown in Fig. 8.16. If the interface is a grounded conductor, then the zero-potential condition is fulfilled by applying an image of the same orders at image point $\mathbf{p}'$ at the same distance below the plane. The image multipole is

$$B'_{n,m} = (-1)^{n+m+1} B_{n,m}. \tag{8.44}$$

Therefore, the resultant potential in the region above the plane becomes

$$\begin{aligned}\phi &= \frac{B_{n,m}}{r_p^{n+1}} \bar{P}_{n,|m|}(\cos\theta_p)\exp(jm\varphi_p) + \frac{B'_{n,m}}{r_{p'}^{n+1}} \bar{P}_{n,|m|}(\cos\theta_{p'})\exp(jm\varphi_{p'}) \\ &= \phi_B + \phi_{B'}\end{aligned} \tag{8.45}$$

where ϕ_B and $\phi_{B'}$ are the potentials due to $B_{n,m}$ and $B'_{n,m}$, respectively.

8.4.2.2 Dielectric Plane

Consider again the configuration in Fig. 8.16, but let the interface be a boundary between two dielectric media having dielectric constant ε_A above and ε_B below the boundary. Two image multipoles, $B'_{n,m}$ at $\mathbf{p}'$ and $B''_{n,m}$ at $\mathbf{p}$, fulfill the boundary

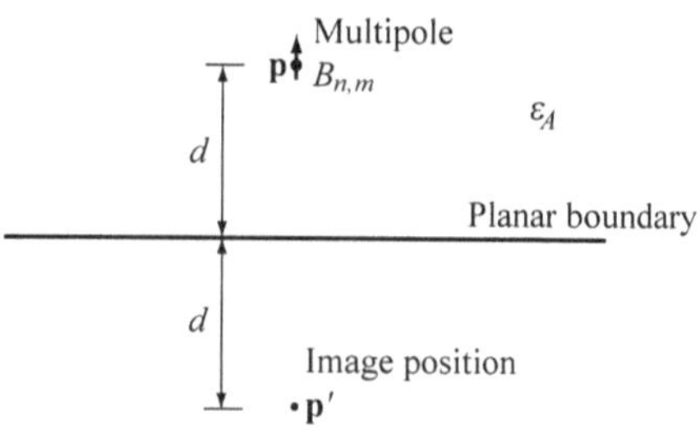

Fig. 8.16 Multipole $B_{n,m}$ at $\mathbf{p}$ above a planar interface and its image at $\mathbf{p}'$

conditions on the planar interface. The image multipoles are related to the source $B_{n,m}$ by

$$B'_{n,m} = (-1)^{n+m+1} P B_{n,m}, \tag{8.46}$$

$$B''_{n,m} = (1 - P) B_{n,m}, \tag{8.47}$$

where $P = (\varepsilon_B - \varepsilon_A)/(\varepsilon_B + \varepsilon_A)$. The resultant potential in the upper region takes the same form as that in Eq. 8.45. In the lower region,

$$\phi = \frac{B''_{n,m}}{r_p^{n+1}} \bar{P}_{n,|m|}(\cos\theta_p)\exp(\mathrm{j}m\varphi_p) = \phi_{B''} \tag{8.48}$$

8.4.2.3 Conducting Sphere

Consider a multipole $B_{n,m}$ of orders (n, m) at point $\mathbf{p}$ and a sphere of radius R centered at point $\mathbf{q}$, as shown in Fig. 8.17. For a spherical object, the multipole potential due to $B_{n,m}$ is re-expanded about $\mathbf{q}$ in the form of a local expansion as

$$\phi = \sum_{j=0}^{\infty}\sum_{k=-j}^{j} M_{j,k} r_q^j \bar{P}_{j,|k|}(\cos\theta_q)\exp(\mathrm{j}k\varphi_q). \tag{8.49}$$

If the sphere is a grounded conductor, the resultant potential outside the sphere is in the form of Eq. 8.38, i.e.,

$$\phi = \sum_{j=0}^{\infty}\sum_{k=-j}^{j}\left(\frac{A_{j,k}}{r_q^{j+1}} + M_{j,k} r_q^j\right)\bar{P}_{j,|k|}(\cos\theta_q)\exp(\mathrm{j}k\varphi_q) \tag{8.50}$$

From Eq. 8.50, the zero-potential condition on the sphere surface at $r_q = R$ is fulfilled if

$$A_{j,k} = -M_{j,k} R^{2j+1} \tag{8.51}$$

for all j and k. In this case, therefore, we may consider that there is an infinite number of image multipoles induced at the sphere center $\mathbf{q}$ with respect to source multipole $B_{n,m}$ at $\mathbf{p}$.

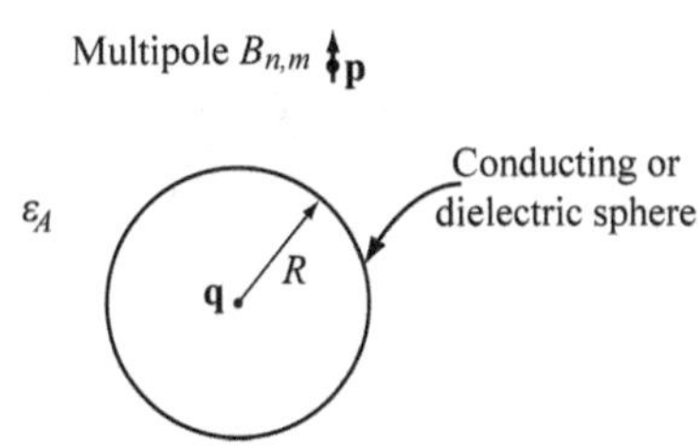

Fig. 8.17 Multipole $B_{n,m}$ at $\mathbf{p}$ located outside a sphere centered at $\mathbf{q}$

Two variations of this configuration are (a) a conducting sphere energized to potential V_0 and (b) an electrically floating conductor. For the former case, we can add $V_0 R$ over the image $A_{0,0}$ in Eq. 8.51. For the latter case, the condition of zero net charge means $A_{0,0} = 0$ and the other $A_{j,k}$ are determined by Eq. 8.51.

8.4.2.4 Dielectric Sphere

For the configuration in Fig. 8.17, if the sphere is a solid dielectric, the potential inside the sphere takes the form of Eq. 8.37,

$$\phi = \sum_{j=0}^{\infty} \sum_{k=-j}^{j} L_{j,k} r_q^j \bar{P}_{j,|k|}(\cos\theta_q) \exp(\mathrm{j}k\varphi_q), \tag{8.52}$$

whereas the potential outside the sphere is still expressed by Eq. 8.50. Denoting the dielectric constant external and internal to the sphere by ε_A and ε_B, respectively, we can determine the unknown potential coefficients from the following relationship:

$$A_{j,k} = \frac{j(\varepsilon_A - \varepsilon_B)}{j(\varepsilon_A + \varepsilon_B) + \varepsilon_A} M_{j,k} R^{2j+1}, \tag{8.53}$$

and

$$L_{j,k} = \frac{(2j+1)\varepsilon_A}{j(\varepsilon_A + \varepsilon_B) + \varepsilon_A} M_{j,k}. \tag{8.54}$$

8.4.2.5 Sphere with a Surface Film

A layered sphere makes a nonhomogeneous object that is sometimes encountered in practice. Figure 8.18 shows the general structure of a layered sphere having core radius R_C and film thickness t. Thus, the overall radius is $R = R_C + t$. The dielectric constants of the external region, the film, and the core are denoted by ε_A, ε_B, and ε_C, respectively. The potential in the core and that outside the sphere are expressed as an expansion about the sphere center **q** in the forms of Eqs. 8.52 and 8.50, respectively. In addition, we write the potential inside the film as:

$$\phi = \sum_{j=0}^{\infty} \sum_{k=-j}^{j} \left(\frac{C_{j,k}}{r_q^{j+1}} + N_{j,k} r_q^j \right) \bar{P}_{j,|k|}(\cos\theta_q) \exp(\mathrm{j}k\phi_q) \tag{8.55}$$

for

$$R_C \le r_q \le R_C + t.$$

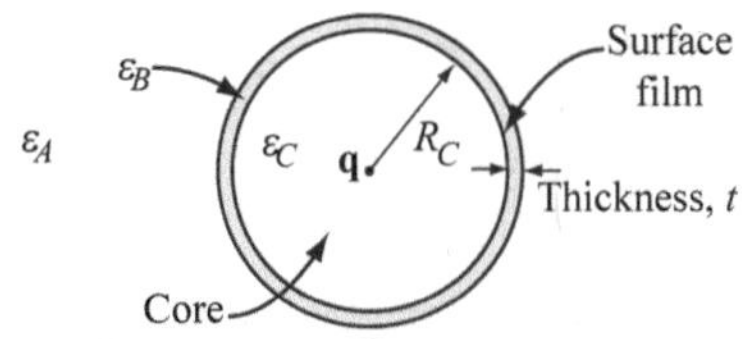

Fig. 8.18 Layered sphere with core radius R_C and film thickness t

If we suppose that the potential coefficients M are given as an applied or background potential, the problem is then to determine the values of A, L, C, and N from M. From the boundary conditions on the core surface, $r_q = R_C$, and those on the outer surface, $r_q = R = R_C + t$, the following relationships can be deduced:

$$\begin{aligned}
A_{j,k} &= M_{j,k}R^{2j+1}\left(\frac{j}{j+1}\right)\left(\frac{K_1 + K_2\varepsilon_{CB} - K_3\varepsilon_{BA} - K_4\varepsilon_{CA}}{K_1 + K_2\varepsilon_{CB} + K_3\varepsilon_{BA} + [j/(j+1)]K_4\varepsilon_{CA}}\right)\\
L_{j,k} &= M_{j,k}\frac{(2j+1)^2}{j+1}\left(\frac{1}{K_1 + K_2\varepsilon_{CB} + K_3\varepsilon_{BA} + [j/(j+1)]K_4\varepsilon_{CA}}\right),\\
C_{j,k} &= L_{j,k}R^{2j+1}\frac{j}{2j+1}(1-\varepsilon_{CB})\\
N_{j,k} &= L_{j,k}\frac{(j+1)+j\varepsilon_{CB}}{2j+1}
\end{aligned} \tag{8.56}$$

where ε_{BA} is defined as the ratio of ε_B to ε_A, and K_1 to K_4 are functions of j and the radius ratio $\zeta = R_C/R$,

$$\begin{aligned}
K_1 &= (j+1) + j\zeta^{2j+1}\\
K_2 &= j(1-\zeta^{2j+1})\\
K_3 &= (j+1)(1-\zeta^{2j+1})\\
K_4 &= j + (j+1)\zeta^{2j+1}.
\end{aligned} \tag{8.57}$$

8.4.3 *Iterative Calculation Procedure*

A general procedure to determine the potential coefficients for a system of particles can be outlined as follows.

1. Begin the calculation with an expression of a given potential condition. If externally applied field $\mathbf{E}_0$ is present in the configuration, write the potential due to $\mathbf{E}_0$ in the form of a local expansion:

$$\phi = \sum_{j=0}^{\infty}\sum_{k=-j}^{j} M_{j,k}r^{j}_{c1}\bar{P}_{j,|k|}(\cos\theta_{c1})\exp(\mathrm{j}k\varphi_{c1}), \tag{8.58}$$

where $\mathbf{c}_1$ is the center of the first particle to which the image scheme is applied. For example, for uniform field $\mathbf{E}_0 = E_0\mathbf{a}_z$ in the z direction,

$$\phi = (z_0 - z_{c1})E_0 - E_0 r_{c1}\bar{P}_{1,0}(\cos\theta_{c1}). \tag{8.59}$$

Thus, $M_{0,0} = (z_0 - z_{c1})E_0$, $M_{1,0} = -E_0$, and the other $M_{j,k}$ are zero. The electric field applied via a pair of parallel plane electrodes is also expressed in a similar way.

For a conducting sphere stressed at potential V_0, the potential ϕ is simply expanded about the sphere center $\mathbf{c}_1$ as

$$\phi = \frac{V_0 R_1}{r_{c1}}, \tag{8.60}$$

where R_1 is the radius of the sphere. Comparing this to Eq. 8.40, it is clear that Eq. 8.60 corresponds to the potential due to a monopole (multipole of zero order) of magnitude $V_0 R_1$ at the sphere center.

2. Express the potentials iteratively through all the particles and electrodes in the configuration, and apply the proper image scheme to satisfy their boundary conditions. For each spherical object, consider the potential due to all the external sources (except that on its own surface), and write it in the form of a local expansion similar to Eq. 8.58. Then, the image multipoles can be determined from Eq. 8.51, 8.53, or 8.56, depending on the type of object.

3. Repeat the previous step until the solutions of all the particles converge.

The following sections illustrate the calculation procedure in more detail.

8.4.4 Two Spherical Particles

Consider a configuration of two spherical particles, P_1 centered at $\mathbf{c}_1$ and P_2 at $\mathbf{c}_2$, under externally applied field $\mathbf{E}_0$, as shown in Fig. 8.19 [10]. Let P_1 be a conductor charged at potential V_1 and P_2 be a solid dielectric having dielectric constant ε_B. We may assume without loss of generality that the particles have the same radius R, and we denote the dielectric constant of the surrounding medium ε_A. The potential can be calculated in the following steps, where subscripts 1 and 2 indicate particles P_1 and P_2, and a superscript (i) indicates the repetition step.

(1) For P_1, write the potential due to $\mathbf{E}_0$ about its center $\mathbf{c}_1$ as

$$\phi_1 = \sum_{j=0}^{\infty}\sum_{k=-j}^{j} M_{1,j,k}^{(1)} r_{c1}^{j} \bar{P}_{j,|k|}(\cos\theta_{c1}) \exp(\mathrm{j}k\,\varphi_{c1}), \tag{8.61}$$

where $M_{1,j,k}^{(1)}$ are determined by $\mathbf{E}_0$ and a reference point of zero potential, as already explained. From $M_{1j,k}$, the image multipoles $A_{1j,k}$ can be computed by using Eq. 8.51 to yield zero potential on the sphere surface. Then, by adding $V_1 R$ to

Fig. 8.19 Conducting particle P_1 and a dielectric particle P_2 under uniform field $\mathbf{E}_0$

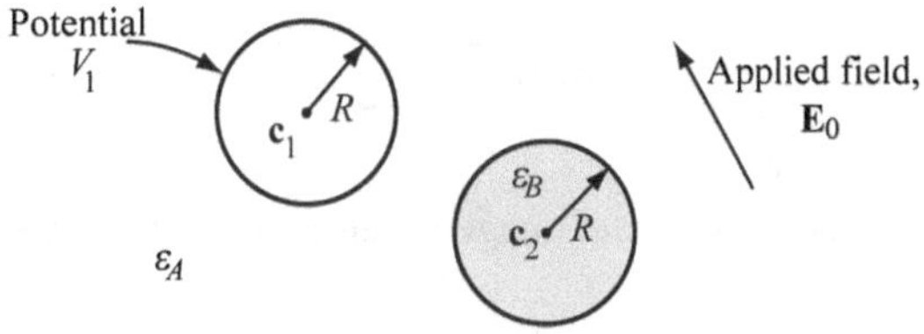

$A_{1,0,0}$, the potential becomes V_1. Therefore, the potential outside P_1 is currently expressed by

$$\phi_1^{(1)} = \sum_{j=0}^{\infty}\sum_{k=-j}^{j}\left(M_{1,j,k}^{(1)} r_{c1}^{j} + \frac{A_{1,j,k}^{(1)}}{r_{c1}^{j+1}}\right)\bar{P}_{j,|k|}(\cos\theta_{c1})\exp(jk\varphi_{c1}). \tag{8.62}$$

(2) For P_2, write the potential in the form of a local expansion about $\mathbf{c}_2$ as

$$\phi_2 = \sum_{j=0}^{\infty}\sum_{k=-j}^{j} M_{2,j,k}^{(1)} r_{c2}^{j}\bar{P}_{j,|k|}(\cos\theta_{c2})\exp(jk\varphi_{c2}), \tag{8.63}$$

where $M_{2j,k}$ represents both the potential due to $\mathbf{E}_0$ and that due to the multipoles $A_{1j,k}$ at $\mathbf{c}_1$ calculated in the previous step. $M_{2j,k}$ is obtained by applying the re-expansion formula in Eq. 8.41. Using the image scheme for a dielectric sphere given in Eq. 8.53, the potential outside P_2 is in the form

$$\phi_2^{(1)} = \sum_{j=0}^{\infty}\sum_{k=-j}^{j}\left(M_{2,j,k}^{(1)} r_{c2}^{j} + \frac{A_{2,j,k}^{(1)}}{r_{c2}^{j+1}}\right)\bar{P}_{j,|k|}(\cos\theta_{c2})\exp(jk\varphi_{c2}). \tag{8.64}$$

(3) Consider P_1 again, and re-expand the multipole potential due to $A_{2j,k}$ in the previous step as

$$\phi_1 = \sum_{j=0}^{\infty}\sum_{k=-j}^{j} M_{1,j,k}^{(2)} r_{c1}^{j}\bar{P}_{j,|k|}(\cos\theta_{c1})\exp(jk\varphi_{c1}), \tag{8.65}$$

where $M_{1,j,k}^{(2)}$ is calculated from $A_{1,j,k}^{(1)}$using Eq. 8.41. Applying Eq. 8.51, we can write the resultant potential as

$$\phi_1^{(2)} = \sum_{j=0}^{\infty}\sum_{k=-j}^{j}\left(M_{1,j,k}^{(2)} r_{c1}^{j} + \frac{A_{1,j,k}^{(2)}}{r_{c1}^{j+1}}\right)\bar{P}_{j,|k|}(\cos\theta_{c1})\exp(jk\varphi_{c1}). \tag{8.66}$$

On the surface of P_1, this potential is zero everywhere; therefore, ${\phi_1}^{(1)} + {\phi_1}^{(2)}$ is equal to V_1.

(4) Consider P_2 again, and re-expand the multipole potential due to $A_{1j,k}$ in the previous step as

$$\phi_2 = \sum_{j=0}^{\infty} \sum_{k=-j}^{j} M_{2,j,k}^{(2)} r_{c2}^{j} \bar{P}_{j,|k|}(\cos\theta_{c2}) \exp(jk\varphi_{c2}). \tag{8.67}$$

Then, using the image scheme in Eq. 8.53, we obtain

$$\phi_2^{(2)} = \sum_{j=0}^{\infty} \sum_{k=-j}^{j} \left(M_{2,j,k}^{(2)} r_{c2}^{j} + \frac{A_{2,j,k}^{(2)}}{r_{c2}^{j+1}} \right) \bar{P}_{j,|k|}(\cos\theta_{c2}) \exp(jk\varphi_{c2}). \tag{8.68}$$

(5) Repeat steps (3) and (4) until the sums

$$\phi_{1,final} = \sum_i \phi_1^{(i)} \tag{8.69}$$

and

$$\phi_{2,final} = \sum_i \phi_2^{(i)} \tag{8.70}$$

converge. After the potential is determined outside the particles, we can use Eq. 8.54 for $M_{2,j,k} = \sum_i M_{2,j,k}^{(i)}$ and write the potential inside the dielectric particle P_2 as

$$\phi_2 = \sum_{j=0}^{\infty} \sum_{k=-j}^{j} L_{2,j,k} r_{c2}^{j} \bar{P}_{j,|k|}(\cos\theta_{c2}) \exp(jk\varphi_{c2}) \tag{8.71}$$

for r_{c2} smaller than or equal to R.

8.4.5 Conducting Particle and a Plane Electrode with a Dielectric Barrier

The configuration is illustrated in Fig. 8.20, in which an electrically floating conductor of radius R is centered at **c** above a grounded plane separated by distance $(h-R)$ [16]. On the plane, there is a dielectric layer (often called a barrier) of thickness t and dielectric constant ε_B. The configuration is subjected to applied field $\mathbf{E}_0$ in the upward vertical direction, and the dielectric constant of the background medium is ε_A.

This configuration is an example of cases in which grouping two or more objects into a compound object may simplify the iterative procedure. In the configuration, the dielectric barrier and the grounded plane are to be treated in the main iteration as a single object. First, we determine the image scheme for source multipole $B_{j,k}$ above the barrier by separation d, as shown in Fig. 8.21a. The scheme is described as follows, where $d^{(i)}$ and $h^{(i)}$ are the distances of a multipole from the upper surface S_A and the lower surface S_B of the barrier in the ith repetition, respectively.

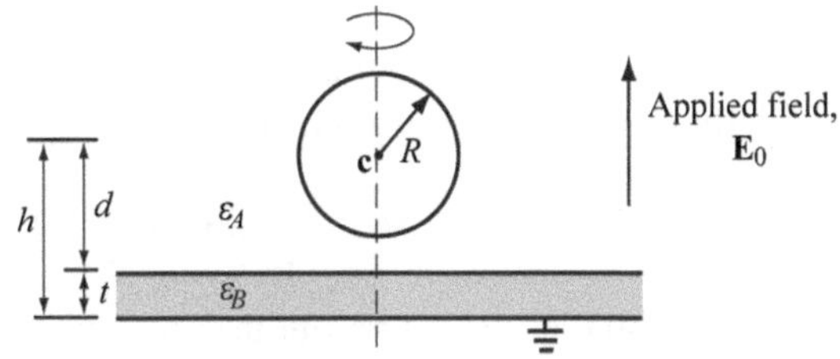

Fig. 8.20 A conducting sphere above a grounded plane with a dielectric barrier

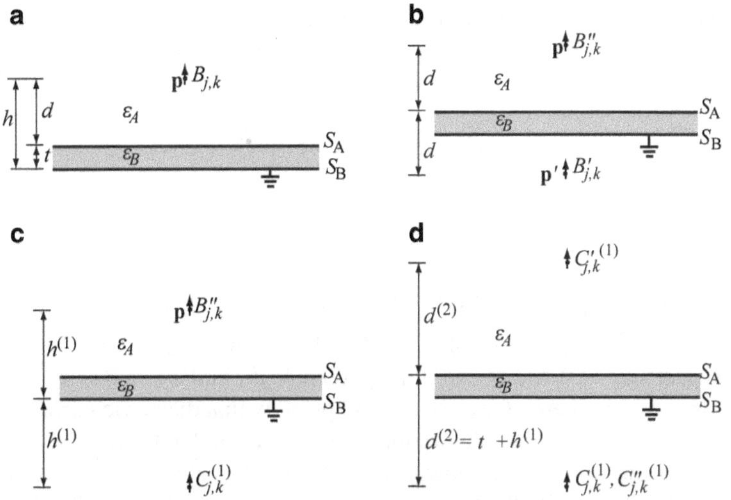

Fig. 8.21 Image scheme for a dielectric barrier on a grounded plane. (**a**) Source multipole $B_{j,k}$ above the upper surface S_A of the barrier at distance d. (**b**) Two image multipoles satisfying the boundary conditions on S_A. (**c**) Image $C_{j,k}{}^{(1)}$ induced by the grounded surface S_B. (**d**) Next images for S_A [16].

(1) Place $B'_{j,k}$ and $B''_{j,k}$ as the images of $B_{j,k}$ (see Fig. 8.21b) using Eqs. 8.46 and 8.47 to satisfy the boundary conditions on S_A. The potential ϕ_A outside the barrier and ϕ_S inside the dielectric barrier are expressed by

$$\begin{aligned}\phi_A &= \phi_B + \phi_{B'} \\ &= \frac{B_{j,k}}{r_p^{j+1}} \bar{P}_{j,|k|}(\cos\theta_p)\exp(jk\varphi_p) + \frac{B'_{j,k}}{r_{p'}^{j+1}} \bar{P}_{j,|k|}(\cos\theta_{p'})\exp(jk\varphi_{p'})\end{aligned} \tag{8.72}$$

$$\phi_S = \phi_{B''} = \frac{B''_{j,k}}{r_p^{j+1}} \bar{P}_{j,|k|}(\cos\theta_p)\exp(jk\varphi_p). \tag{8.73}$$

(2) Insert multipole ${C_{j,k}}^{(1)}$ at distance $h^{(1)} = h$ below the boundary plane as the image of $B''_{j,k}$ with respect to the conducting surface S_B (see Fig. 8.21c). The image is determined by using Eq. 8.44, i.e.,

$$C_{j,k}^{(1)} = (-1)^{j+k+1} B''_{j,k}. \tag{8.74}$$

The potential inside the barrier then becomes

$$\phi_S = \phi_{B''} + \phi_{C^{(1)}}. \tag{8.75}$$

(3) Place $C'_{j,k}(1)$ and $C''_{j,k}(1)$as the images of $C_{j,k}^{(1)}$ induced by the interface S_A, as shown in Fig. 8.21d. Equations 8.46 and 8.47 are used again, but the constant P in this step is equal to $(\varepsilon_A - \varepsilon_B)/(\varepsilon_A + \varepsilon_B)$, different from that used in the first step. This is because $C_{j,k}^{(1)}$ contributes to the potential in the dielectric barrier. After this step, the sources of the potential for each region become

ε_A region: $B_{j,k}, B'_{j,k},$ and $C''_{j,k}(1)$;
ε_B region: $B''_{j,k}, C_{j,k}^{(1)}$, and $C'_{j,k}(1)$.

(4) Because $C'_{j,k}(1)$ perturbs the potential in the barrier, its image $C_{j,k}^{(2)} = (-1)^{j+k+1} C'_{j,k}(1)$ with respect to the surface S_B is applied at $h^{(2)} = d^{(2)} + t$.

(5) Repeat steps (3) and (4) until the potentials ϕ_A and ϕ_S converge.

With the image scheme for a conducting plane with a dielectric barrier given as above, the potential in the configuration of Fig. 8.20 can be determined by following the calculation procedure shown in Fig. 8.21. The calculation begins with the solution of the electrically floating conducting sphere under the field $\mathbf{E}_0$. Then, a set of the image multipoles induced by the compound object (the plane and the barrier) is calculated. Next, the potential due to $B'_{j,k}$ and all $C''_{j,k}(i)$ of the current iteration is re-expanded about $\mathbf{c}$. As a result, it is written in the form

$$\phi_{B'} + \sum_i \phi_{C''(i)} = \sum_{j=0}^{\infty} \sum_{k=-j}^{j} M_{j,k} r_c^j \bar{P}_{j,|k|}(\cos\theta_c)\exp(\mathrm{j}k\varphi_c). \tag{8.76}$$

In the next step, the image scheme for a floating conducting sphere is applied to compute $B_{j,k}$ from $M_{j,k}$ for the iteration. These calculation procedures from (1) to (5) are summarized as a flowchart in Fig. 8.22.

Note that, as explained in Section 8.4.1, the equations become much simpler for axisymmetric cases. In the configuration of Fig. 8.20, all the potential coefficients associated with nonzero k vanish. For example, Eq. 8.76 reduces to

$$\phi_{B'} + \sum_i \phi_{C''(i)} = \sum_{j=0}^{\infty} M_{j,0} r_c^j P_j(\cos\theta_c). \tag{8.77}$$

The normalized Legendre function becomes the simpler $P_j(\cos\theta_c)$ in Eq. 8.77.

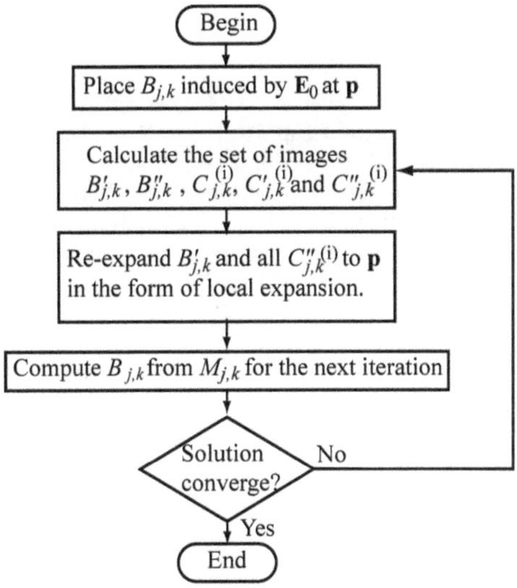

Fig. 8.22 Calculation procedure for a conducting sphere and a conducting plane with a dielectric barrier, as shown in Fig. 8.20 [16]. © 2006 IEEE

References

1. Takuma, T., Kouno, T., Matsuda, H.: Field behavior near singular points in composite dielectric arrangements. IEEE Trans. Electr. Insul. **13**(6), 426–435 (1978)
2. Takuma, T., Kouno, T., Matsuba, H., Watanabe, T., Kawamoto, T.: On the field distribution at the boundary surface of two dielectrics having a straight section. CRIEPI (Central Research Institute of Electric Power Industry) Research Report No.176002 (1976) (in Japanese)
3. Takuma, T., Kawamoto, T.: Field behavior at a void edge in a solid dielectric. The 1989 Annual Meet. Rec. of IEEJ (Inst. Electr. Eng. Japan) No. 22 (1989) (in Japanese)
4. Takuma, T., Kawamoto, T.: Field enhancement at a triple junction in arrangements consisting of three media. IEEE Trans. Dielectr. Electr. Insul. **14**(3), 566–571 (2007)
5. Takuma, T., Kawamoto, T.: Field intensification near various points of contact with a zero contact angle between a solid dielectric and an electrode. IEEE Trans. Power Appar. Syst. **103** (9), 2486–2494 (1984)
6. Takuma, T.: Field behavior at a triple junction in composite dielectric arrangements. IEEE Trans. Electr. Insul. **26**(3), 500–509 (1991)
7. Takuma, T., Kawamoto, T.: Field behavior when a solid dielectric is in contact with an electrode. Papers of Techn. Meet. on Electrical Discharge IEEJ (Inst. Electr. Eng. Japan): ED-83-14 and ED-83-50 (1983) (in Japanese)
8. Techaumnat, B., Takuma, T.: Electric field and force on a conducting sphere in contact with a dielectric solid. J. Electrost. **64**(3–4), 165–175 (2006)
9. Washizu, M., Jones, T.B.: Dielectrophoretic interaction of two spherical particles calculated by equivalent multipole-moment method. IEEE Trans. Ind. Appl. **32**(2), 233–242 (1996)
10. Techaumnat, B., Eua-arporn, B., Takuma, T.: Calculation of electric field and dielectrophoretic force on spherical particles in chain. J. Appl. Phys. **95**(3), 1586–1593 (2004)
11. Rokhlin, V.: Rapid solution of integral equations of classical potential theory. J. Comp. Phys. **60**(2), 187–207 (1983)
12. Hamada, S., Yamamoto, O., Kobayashi, T.: Analysis of electric field induced by ELF magnetic field utilizing generalized equivalent multipole-moment method. Electr. Eng. Jpn. **156**(2), 1–14 (2006)

13. Kitano, M., Hamada, S., Kobayashi, T.: Analytical formula of induced electric fields in a spherical conductor by an ELF dipole magnetic field source. Electr. Eng. Jpn. **166**(3), 8–17 (2008)
14. Takuma, T., Hamada, S.: Fundamentals and applications of numerical calculation methods of electric fields, Chap. 10. Tokyo Denki University Press (2006) (in Japanese)
15. Greengard, L., Rokhlin, V.: A new version of the fast multipole method for the Laplace equation in three dimensions. Acta Numer. **6**, 229–269 (1997)
16. Techaumnat, B., Takuma, T.: Analysis of the electric field and force in an arrangement of a conducting sphere and a plane electrode with a dielectric barrier. IEEE Trans. Dielectr. Electr. Insul. **13**(1), 336–344 (2006)

Chapter 9
Numerical Calculation Methods

Introduction

Numerical calculation methods are those which use discrete values or simple functions at small intervals or over small elements as a substitute for continuously changing field values. The first numerical calculation of electric fields is attributed to the work of J.C. Maxwell in 1887. He computed the approximate value of the capacitance for an isolated conducting square by dividing the square (and consequently the charge) into 36 (i.e., 6×6) small squares.

In the sections that follow the general remarks, we principally explain the so-called boundary-dividing methods, which are considered more suitable for analyzing electric fields in composite dielectrics than the domain-dividing methods.

9.1 General Remarks

Although the basic features of various numerical methods are already well known, we here briefly summarize them in the explanations below [1–3]. The focus lies in the characteristic differences of various methods when they are applied to calculations related to composite dielectric configurations.

Numerical calculation methods of electric or magnetic fields are fundamentally divided into two categories, i.e., domain-dividing methods and boundary-dividing methods. They are also referred to, respectively, as differential equation methods and integral equation methods, according to the forms of the respective system equations. In the former category, the whole domain of interest is subdivided into small elements and the resulting system equation consists of space potentials as unknowns. The domain-dividing methods include the finite difference method (FDM), the finite element method (FEM), and the Monte-Carlo method (MCM). In boundary-dividing methods for electric field calculations, only boundaries (conductor surfaces and dielectric interfaces) are subdivided, and a system equation is formulated as an integral equation for charges or charge densities (also with surface

T. Takuma and B. Techaumnat, *Electric Fields in Composite Dielectrics and their Applications*, Power Systems, DOI 10.1007/978-90-481-9392-9_9,

potentials in the boundary element method) as unknowns. The boundary-dividing methods include the charge simulation method (CSM), the integral equation method (IEM), the surface charge method (SCM), the surface charge simulation method (SCSM), and the boundary element method (BEM).

These terms for the boundary-dividing methods are sometimes confused. In this book, we refer to CSM, SCM, and BEM according to the following definitions. CSM utilizes a number of fictitious charges to equivalently express the electric field in a problem. To simulate conductors, these charges are placed inside the conductors and not directly on their surfaces where the true charge is located. For dielectric interfaces, these charges are not placed on the interface where the polarization charge is located, but inside both dielectrics at some distance from the interface. Both SCM and BEM divide conductor surfaces or dielectric interfaces into a number of surface elements (often called patches), but the unknowns are solely the true and polarization charges on the patches in SCM, while they are potentials and normal components of the electric field (or electric flux density) in BEM. The unknowns are usually different between the two methods, except for monodielectric configurations where the boundaries consist only of conductor surfaces.

Table 9.1 briefly compares the basic features of the four principal methods used for calculating electric fields. It is to be noted in the table that the maximum

Table 9.1 Comparison of principal numerical calculation methods for electric fields [1–3]

	Domain-dividing methods		Boundary-dividing method	
	FDM	FEM	CSM	SCM, BEM
Unknowns	Potentials at lattice (grid) points	Potentials at nodes	Fictitious charges	Charge densities for SCM, potentials and field strengths for BEM
Max. number of unknowns	$\approx 10^7$		$\approx 5 \times 10^4$	
Coefficient matrix[a]	Sparse		Full or dense	
How to obtain field values	(Potential difference)/distance or numerical differentiation of potentials		Analytical expressions for fields caused by charges	Numerical integral of fields caused by charge densities (or analytical expressions for planar charges)
Applicability and other features	Applicable to any problem, including nonlinear cases	Applicable to any problem, including nonlinear cases	Applicable to Laplacian fields, in particular, suitable for 2D and AS conditions	Applicable mainly to Laplacian fields, but more general than CSM
	Easy to subdivide domains	More flexible than FDM, but more complicated programming and input data	Needs experience and intuition to adopt proper positions of charges and contour points	Often troublesome in numerical integration when a computation point coincides with a source (charge)
	Difficult to handle complicated or curved boundaries	Suitable for complex, intricate problems	Difficult to deal with thin materials	

[a]Of a system equation

available number of unknowns has remarkably increased with the recent development of computer capacity.

The boundary-dividing methods suffer from an inherent disadvantage that they cannot handle most nonlinear problems. In such nonlinear cases, the division of a whole domain is usually inevitable to express the entire field. However, compared with the domain-dividing methods, the boundary-dividing methods possess the following advantages:

(a) Easier subdivision due to the curtailed dimensions,
(b) Higher accuracy of field values computed without differentiating potentials, and
(c) Easier treatment (inclusion) of an infinite domain

For the analysis of electric fields in composite dielectrics, in particular, field enhancement near a contact point, the boundary-dividing methods are more appropriate mainly due to calculation accuracy. In fact, the computation results in this book are exclusively derived using boundary-dividing methods, although a few results generated by domain-dividing methods are referred to and commented on.

9.2 Charge Simulation Method (CSM)

9.2.1 Basic Principle

The charge simulation method (originally *Ersatzladungsmethode* in German) was founded by H. Steinbigler in his dissertation submitted to the Technical University Munich in 1969 [4, 5]. The principle of the ordinary CSM is to approximate the actual field under study with a field which is formed by a discrete set of fictitious charges placed inside conductors (electrodes). The type (form) and position of these charges are predetermined, but their magnitude is unknown and is determined from the boundary conditions. Explicit charge forms are usually applied as fictitious charges which enable us to use analytical expressions for both the potential and field strength. Typically utilized are infinite line charges in two-dimensional (2D) fields, while point, line, and ring charges are used in axisymmetric (AS) cases. They are illustrated, respectively, in Figs. 9.1a and b.

The potential resulting from a set of fictitious charges of magnitude $Q(j)$ is given at point i as

$$\phi(i) = \sum_{j=1}^{N} P(i, j)Q(j), \tag{9.1}$$

where N is the number of fictitious charges and $P(i, j)$, called the potential coefficient, means the potential at point i caused by a unit charge of $Q(j)$. It depends only on the type (form) of the charge and the relative position or distance between i and

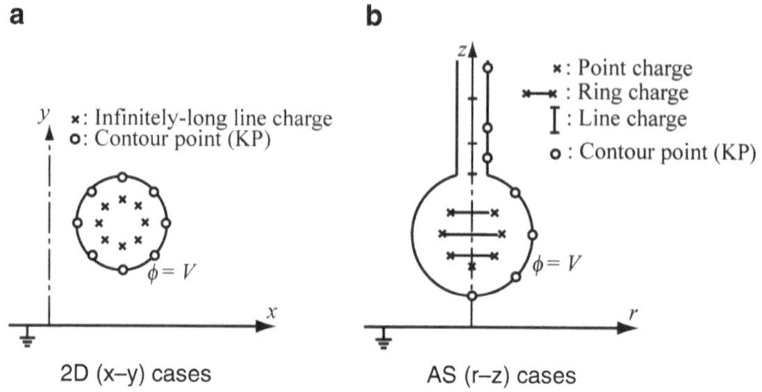

Fig. 9.1 Fictitious charges and contour points (shown schematically) used in CSM.

charge $Q(j)$. Because each $P(i, j)$ satisfies Laplace's equation, $\phi(i)$ is a justified solution if it satisfies the boundary conditions. Imposing the boundary condition of a constant potential at a set of appropriately chosen points (called contour points, or KPs, after the German word *Konturpunkt*) on the conductor surface leads to a simultaneous linear equation system for solving $Q(j)$ as unknowns

$$\sum_{j=1}^{N} P(i,j)Q(j) = V_i, \tag{9.2}$$

where V_i is a potential or an applied voltage on the conductor. The number of contour points is usually chosen to be equal to N. Once the $Q(j)$ values are thus determined from Eq. 9.2, the potential can be given at any point in the region by applying Eq. 9.1. On the other hand, the field strength at point i is given by

$$\mathbf{E}(i) = \sum_{j=1}^{N} \mathbf{F}(i,\ j)Q(j). \tag{9.3}$$

An element of the coefficient matrix $\mathbf{F}(i, j)$ means the field strength at point i caused by a unit charge of $Q(j)$.

9.2.2 Composite Dielectric Cases

In configurations consisting of two or more dielectrics, we substitute fictitious charges placed inside each of the dielectrics to simulate polarization charges on dielectric interfaces, as schematically shown in Fig. 9.2. In the case of two dielectrics, A (dielectric constant ε_A) and B (ε_B), three sets of fictitious charges, $Q(i)$, $Q_A(i)$, and $Q_B(i)$, are located, respectively, inside the conductors (electrodes) and inside dielectrics A and B.

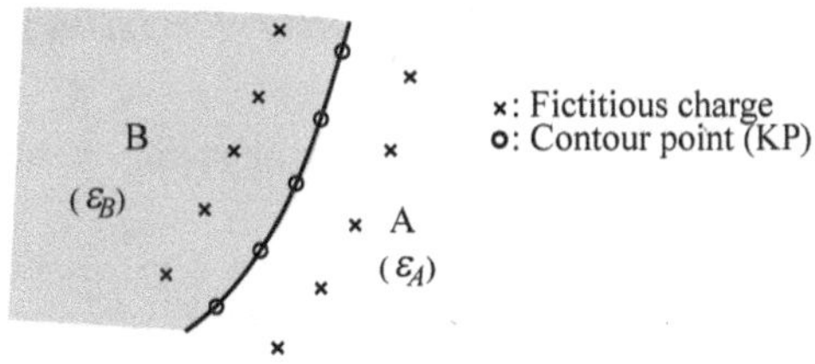

Fig. 9.2 Arrangement of fictitious charges and contour points (KPs) to simulate polarization charge at a dielectric interface

The boundary conditions on the conductor surfaces and dielectric interfaces are explained in Section 1.2.2. Since two charges correspond to each KP on the interfaces, dual boundary conditions are imposed on each KP: the continuity of the potential and the normal component of the electric flux density, i.e.,

$$\phi_A(i) = \phi_B(i),$$

and

$$\varepsilon_A E_{nA}(i) = \varepsilon_B E_{nB}(i). \tag{9.4}$$

In the usual CSM, the field in dielectric A is represented by the effect of $Q(i)$ and $Q_B(i)$, while that in B is given by the effect of $Q(i)$ and $Q_A(i)$. The other procedures are carried out in a similar manner as in the monodielectric cases explained above in Section 9.2.1, which leads to Eqs. 9.1 and 9.3 also for composite dielectric configurations.

9.2.3 *β-Method: CSM Using Fictitious Charges Inside Surrounding Boundaries Only*

When the values of the two dielectric constants ε_A and ε_B are very different (e.g., $\varepsilon_A << \varepsilon_B$), conventional CSM may result in a significantly large relative error in the region with the larger constant ε_B. The error becomes more noticeable for fields where the effect of conductivity is predominant over that of the dielectric constant, because conductivity may differ by orders of magnitude among materials or media. The reason for the error is that the field in dielectric B is expressed by the combined effect of two discretely located charge groups, $Q(i)$ inside the conductor and $Q_A(i)$ inside A. In this situation, the effect of $Q(i)$ and $Q_A(i)$ must act in opposite directions to counteract each other to ensure a weaker field in B, but this condition cannot be easily realized with a finite number of discrete charges.

P. Weiss mentioned in his dissertation a variation of the CSM which simulates the field only with the effect of fictitious charges situated near surrounding boundaries [6]. T. Takuma et al. furthermore studied the applicability of this method (called the β-method) for a model configuration of a dielectric sphere existing under a uniform field with and without conductivity, and confirmed that the β-method maintains good accuracy even for cases in which the conventional CSM fails [7].

9.2.4 *Mixed (Capacitive-Resistive) Fields*

9.2.4.1 Complex CSM for ac steady fields

Calculation with complex variables, or phasor notation, is explained in Section 1.3.1. When all the waveforms of voltage sources in a system are sinusoidal with the same angular frequency ω, we can conveniently express the instantaneous field directly with complex fictitious charges in the CSM. This is permitted since the magnitude of fictitious charges in an ac steady state also changes sinusoidally with the same angular frequency ω. [8, 9]

Thus, in the CSM, instead of Eqs. 9.1 and 9.2, the system equations become

$$\dot{\phi}(i) = \sum_{j=1}^{N} P(i,j)\dot{Q}(j) \tag{9.5}$$

at point i in the region, and

$$\sum_{j=1}^{N} P(i,j)\dot{Q}(j) = \dot{V}_i \tag{9.6}$$

at contour point i on the electrode surfaces. Equation 9.6 constitutes a set of simultaneous linear equations for the complex unknowns $\dot{Q}(j)$ with real coefficients.

9.2.4.2 CSM for fields including volume conduction

The basic equations for fields including volume conduction are also explained in Section 1.3.1. In a region with dielectric constant ε and with volume conductivity σ, electric field $\mathbf{E}$ is formulated for an ac steady-state field of angular frequency ω with a complex field as Eq. 1.14,

$$\mathrm{div}\left\{(\sigma + \mathrm{j}\omega\varepsilon\varepsilon_0)\dot{\mathbf{E}}\right\} = 0. \tag{9.7}$$

This equation means that the field can be expressed simply by substituting the following complex dielectric constant for ε in ordinary composite dielectric fields without conduction

$$\dot{\varepsilon} = \varepsilon + \frac{\sigma}{\mathrm{j}\omega\varepsilon_0}. \tag{9.8}$$

If both ε and σ are constant in a medium, Eq. 9.8 again leads to div $\dot{\mathbf{E}} = 0$, i.e., Laplace's equation, inside the medium.

In the case of two dielectrics with volume conductivity, i.e., A (dielectric constant ε_A, volume conductivity σ_A) and B (ε_B, σ_B), as shown in Fig. 9.3 (same as Fig. 1.1),

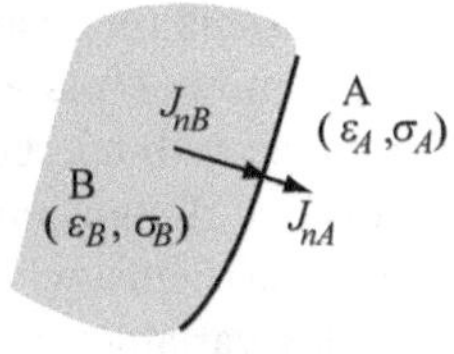

Fig. 9.3 Interface of dielectrics with volume conduction. J_{nA} and J_{nB} are the normal component of current density flowing in A and B

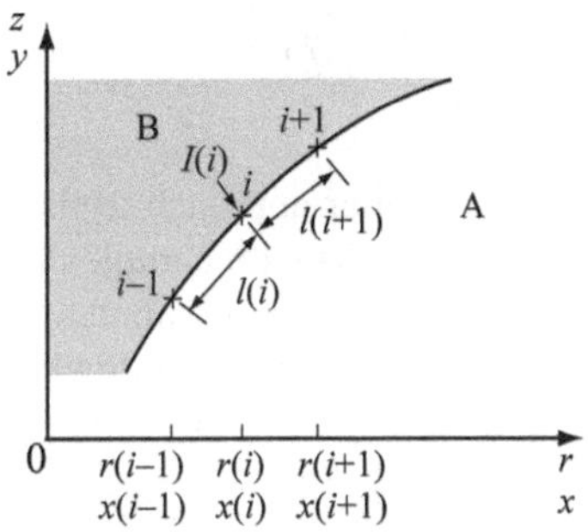

Fig. 9.4 Dielectric interface with surface conduction in AS (r, z) or 2D (x, y) conditions

the continuity condition of the normal component of the electric flux density at the interface results in the following equation, as described in Section 1.3.2,

$$\left(\varepsilon_A + \frac{\sigma_A}{j\omega\varepsilon_0}\right)\dot{E}_{nA} = \left(\varepsilon_B + \frac{\sigma_B}{j\omega\varepsilon_0}\right)\dot{E}_{nB}. \tag{9.9}$$

Equation 9.9 is the same as the boundary condition (Eq. 9.4) for the interfaces of usual dielectrics without conduction where ε is replaced with the complex dielectric constant of Eq. 9.8.

As explained above in Section 9.2.3, when mixed fields are formed by regions with remarkably different conductivities, the conventional CSM may result in a significantly large relative error in the region with the larger conductivity. The β-method mentioned there is effective for computing these fields where the field in each medium is expressed only by fictitious charges existing near the surrounding boundaries.

9.2.4.3 CSM for fields including surface conduction

For configurations with surface conduction, the basic concept is to express the true surface charge (density q_s) at point i on the interface in relation to current $I(i)$ flowing along the interface, which is schematically illustrated in Fig. 9.4. The resulting equation is

$$q_s(i) = \frac{1}{S(i)}\int_0^t I(i)dt = \frac{1}{S(i)}\int_0^t \left\{\frac{\phi(i-1)-\phi(i)}{R(i)} - \frac{\phi(i)-\phi(i+1)}{R(i+1)}\right\}dt, \tag{9.10}$$

where $R(i)$ is the resistance between two neighboring contour points (KPs), i and $(i-1)$, while $S(i)$ is an infinitesimal interface area at point i. In this equation, the current path should be treated as one-dimensional, although the original configuration is two-dimensional (2D) or axisymmetric (AS).

For ac fields with angular frequency ω, this equation can be treated with complex variables as

$$\dot{q}_s(i) = \frac{1}{j\omega S(i)}\left\{\frac{\dot{\phi}(i-1)-\dot{\phi}(i)}{R(i)} - \frac{\dot{\phi}(i)-\dot{\phi}(i+1)}{R(i+1)}\right\}. \tag{9.11}$$

In a similar approach to that for mixed fields with volume conduction, we can apply the CSM with complex fictitious charges to compute fields including surface conduction. In the computation process, Eq. 9.11 is used to give a true surface charge density for the following boundary condition

$$\varepsilon_A \dot{E}_{nA}(i) - \varepsilon_B \dot{E}_{nB}(i) = \dot{q}_s(i). \tag{9.12}$$

In 2D arrangements, we can also utilize the following expression for the true surface charge density q_s instead of Eq. 9.10,

$$q_s(i) = \frac{2}{l(i)+l(i+1)}\int_0^t [\sigma_s(i)E_t(i) - \sigma_s(i+1)E_t(i+1)]dt. \tag{9.13}$$

In this equation, $l(i)$ is the distance between point i and $(i–1)$, $\sigma_s(i)$ is the surface conductivity, and $E_t(i)$ is the average tangential field strength along the interface. These values respectively belong to point i.

9.2.5 Example of Boundary Division

Special attention must be paid when dealing with the field singularity near a point of contact in composite dielectrics having cross-sectionally straight interfaces. The contour points (KPs) and fictitious charges should be arranged so as to conform to the field behavior in the vicinity of the contact point. T. Takuma et al. apply the following procedure for a contact with angle α [10], as schematically shown in Fig. 9.5.

In 2D $x-y$ coordinates, the x-axis is taken along a straight (linear) boundary, i.e., the conductor surface or dielectric interface, where the position of a contour point

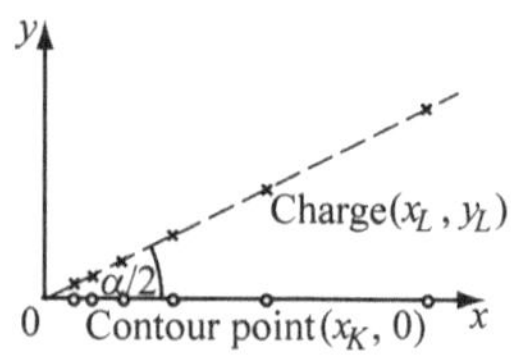

Fig. 9.5 Arrangement of contour points (KPs) and fictitious charges near a contact point formed with linear boundaries

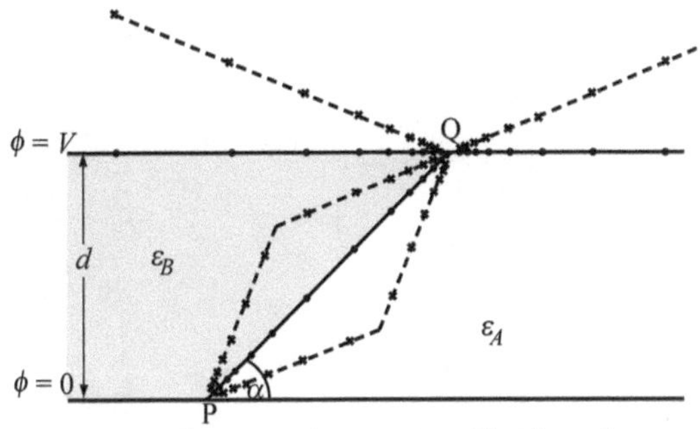

Fig. 9.6 Exemplary arrangement (schematically shown) of contour points (KPs) and fictitious charges used for calculating the field behavior in Fig. 2.6 [10]. © 1978 IEEE

(KP) is $x_K(j)$ at point j, while the position of the corresponding fictitious charge (*Ladung* in German) is denoted by $x_L(j)$ and $y_L(j)$.

$$x_L(j) = x_K(j),\ y_L(j) = \tan(\alpha/2)x_L(j) \tag{9.14}$$

Based on their experiences of such calculations, the authors furthermore set the following relation in the two surrounding KPs:

$$y_L(j) = 1.2[x_K(j) - x_K(j-1)]. \tag{9.15}$$

These conditions ensure that the arrangement of KPs and fictitious charges become denser in a geometric series as the contact point is approached. Figure 9.6 demonstrates an example of such arrangements, as applied to the linear, oblique dielectric interface between two parallel plane electrodes of Fig. 2.6.

A similar arrangement procedure is also applied for a boundary which is not straight near a contact point, as encountered in the contact of a rounded solid material edge with a planar one. The function $x_K(j)$ in Eqs. 9.14 and 9.15 gives the position of a KP at point j as the distance from the contact point along the curved boundary, while $x_L(j)$ denotes the distance of a corresponding fictitious charge from the KP on the line perpendicular to the boundary there. An exemplary arrangement is shown in Fig. 9.7 for a circular (quadrantal) cross-sectional edge of a dielectric interface in the contact of three dielectrics, i.e., the configuration shown in Fig. 4.7.

9.3 Surface Charge Method (SCM)

9.3.1 Basic Principle

The electrostatic field is adequately expressed solely by the effect of true charge on conductor surfaces and polarization charge on dielectric interfaces. As briefly mentioned in Section 9.1, the surface charge method (SCM) divides boundaries (conductor surfaces and dielectric interfaces) into a number of small elements (patches), and utilizes the charge magnitude or charge density on each patch as

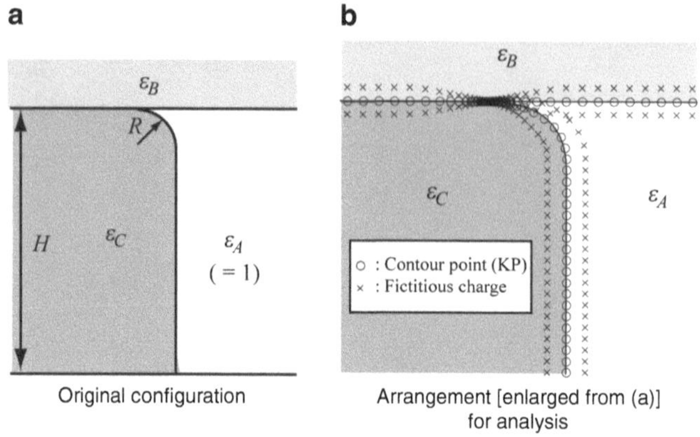

Fig. 9.7 Example of arrangement of contour points (KPs) and fictitious charges used for analyzing configurations consisting of three dielectrics, as discussed in Section 4.3.2 [11]

an unknown to be solved. In this context, the SCM is considered a more straightforward method than the charge simulation method (CSM) explained in Section 9.2, which uses fictitious charges not existing on the boundaries. It should also be noted that on composite dielectric interfaces, the CSM necessitates twice as many unknowns as the SCM because two fictitious charges are placed there corresponding to each contour point in the CSM. Because the SCM was rarely applied for the calculations discussed in the preceding chapters of this book, we only briefly mention the basic, qualitative features of the method together with some recently improved techniques.

The SCM simply integrates fractional field contributions from all the charges on the boundary surfaces as

$$\phi = \frac{1}{4\pi\varepsilon_0}\int \frac{q_s ds}{l},\ \mathbf{E} = -\frac{1}{4\pi\varepsilon_0}\int \nabla\left(\frac{1}{l}\right) q_s ds \tag{9.16}$$

In these equations, ϕ and $\mathbf{E}$ are the potential and electric field at an arbitrary point P, q_s is the charge density on boundary surface element ds, and l is the distance between P and q_s. For numerical calculation, an integral is converted to the fractional summation of contributions from the subdivided elements (patches), with a boundary condition imposed at an appropriate point on each element. This point-matching condition corresponds to the application of the Dirac's δ function in the method of moments.

The SCM is basically divided into the following two categories:

(a) SCM which utilizes analytical expressions of potentials and electric fields without resorting to numerical integration (hereafter called SCM-A), and
(b) SCM which gives potentials and fields by numerical integrals (SCM-B).

SCM-B entails complicated computation techniques on the boundaries. This is because both the electric potential and the field strength become singular when calculating point P coincides with source charge q_s, and thus the distance l becomes zero in Eq. 9.16. On the other hand, SCM-A does not need the troublesome treatment of such singularities, but curved surfaces must be simulated as polyhedrons because analytical expressions are available only for planar charges.

Although the SCM has already been sufficiently developed in monodielectric configurations where boundaries consist of conductor surfaces only, the calculation is still troublesome in composite dielectrics. This is mainly because at the edges of the dielectric interface we cannot easily impose the boundary condition, i.e., the continuity of the normal component of the electric flux density, as the field is infinitely high there. Second, a field singularity may arise at a contact point between a dielectric interface and a conductor surface or another interface. This phenomenon, often called the triple-junction effect, is fully explained in Chapter 2 and elsewhere.

9.3.2 *Some Improvements*

The following techniques have remarkably improved the accuracy of the method in calculating electric fields compared with the simple, unsophisticated SCM.

9.3.2.1 Simulation of Rounded Surface Profiles

Most parts of high-voltage equipment are composed of rounded surfaces to suppress possible discharge (or corona) inception. This makes the SCM that applies to curved surfaces (SCM-B) more appropriate in most cases than SCM-A, which deals with polyhedral surfaces. H. Singer's group proposed the use of bi-cubic spline functions for arbitrary 3D surfaces in the boundary element method (BEM) [12]. S. Hamada's group applied cubic triangular and quadrilateral Bézier patches (surface elements) to represent arbitrary surface shapes, which are respectively equivalent to Zienkiewicz's and cubic serendipity patches [13]. The application of rational quartic triangular Bézier patches is effective in improving the accuracy of the calculated field when we consider a sphere or its octants because it expresses the octant of a sphere with neither shape error nor degeneration of nodes [14]. S. Hamada's group furthermore introduced a quadrilateral Gregory patch and extended it to a new triangular version based on the Bézier patch [15]. This method permits the smooth connection of all patches along their boundary curves, as shown in Fig. 9.8, and enables us to utilize higher-order trial functions of surface charge density q_s in the SCM when point-matching is applied.

9.3.2.2 Expression of Surface Charge Density

Higher accuracy can be attained when we use higher-order functions relative to the space coordinates for simulating q_s. Most conventional SCM calculations use first-order functions, at the highest, for monodielectric configurations and zero-order functions for composite dielectrics due to the wedge-like connection of neighboring patches. H. Singer's group applied the following linear charge distribution on a quadrilateral including a uv-term on the surface coordinate (u, v) [12]

$$q_s = C_0 + C_1 u + C_2 v + C_3 uv. \tag{9.17}$$

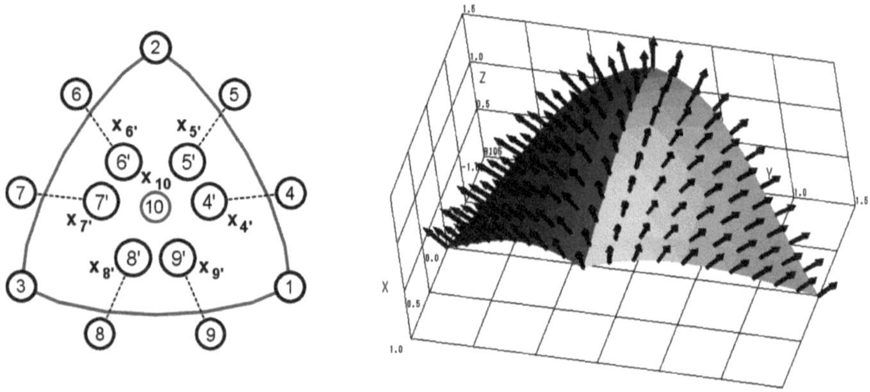

Fig. 9.8 Proposed triangular patch with variable interior control points (Nos. 4′–9′) [15]

For SCM-A, analytical expressions were derived up to the second and third orders of trial functions for planar charges [16]. In SCM-B with curved surfaces, the above-mentioned restriction of representing q_s can be removed by the smooth connection of patches, as explained above in 9.3.2.1.

9.3.2.3 Formulation of Boundary Conditions

Instead of the conventional point-matching algorithm, S. Hamada's group proposed a method to fulfill the continuity condition of the electric flux density averaged on each subdivided patch [17]. This technique also makes possible the higher-order representation of q_s by applying an adequately weighted residual method to the boundary conditions.

9.4 Boundary Element Method (BEM)

9.4.1 Basic Equations

We explain the boundary element method (BEM) based on the boundary integral method derived from Green's theorem [18]. The direct BEM expresses potential $\phi(\mathbf{r})$ at point $\mathbf{r}$ from two integrals on the boundary Γ enclosing homogeneous region Ω_A, in which the point $\mathbf{r}$ is located.

$$c\phi(\mathbf{r}) = -\int_{\Gamma} w(\mathbf{r}, \mathbf{r}_s) E_n \mathrm{d}S - \int_{\Gamma} \frac{\partial w(\mathbf{r}, \mathbf{r}_s)}{\partial n} \phi \mathrm{d}S, \tag{9.18}$$

where $w(\mathbf{r}, \mathbf{r}_s)$ is a fundamental solution, E_n is an outward normal electric field, and the integrals are to be taken for $\mathbf{r}_s$. The constant c is the ratio of the corresponding

solid angle inside Ω_A at $\mathbf{r}$ to 4π (the total solid angle). That is, $c = 1$ when $\mathbf{r}$ is inside Ω_A, and $c = 1/2$ when $\mathbf{r}$ is on a smooth boundary. The fundamental solution, w, can be expressed as follows.
Two-dimensional (2D) case:

$$w(\mathbf{r}, \mathbf{r}_s) = \frac{1}{2\pi} \ln \frac{1}{\|\mathbf{r} - \mathbf{r}_s\|}, \tag{9.19}$$

where the second norm $\|\mathbf{r} - \mathbf{r}_s\|$ is the distance between $\mathbf{r}$ and $\mathbf{r}_s$.
Three-dimensional (3D) case:

$$w(\mathbf{r}, \mathbf{r}_s) = \frac{1}{4\pi\|\mathbf{r} - \mathbf{r}_s\|}. \tag{9.20}$$

Axisymmetric (AS) case:
Let the polar coordinates of $\mathbf{r}$ and $\mathbf{r}_s$ be (ρ, z) and (ρ_s, z_s), respectively, then

$$w(\mathbf{r}, \mathbf{r}_s) = \frac{K_m}{\sqrt{a + b}}, \tag{9.21}$$

where K_m is an elliptic integral of the first kind, and

$$m = \sqrt{2b/(a + b)}, \tag{9.22}$$

$$a = \rho^2 + \rho_s^2 + (z - z_s)^2, \tag{9.23}$$

$$b = \rho\rho_s. \tag{9.24}$$

Equation 9.18 indicates that once we know the potential and the electric field normal to the boundary Γ, we can determine the potential at any point inside Ω. The boundary Γ is discretized into small elements or patches, and the potential and the field on each element are interpolated by

$$\phi = \sum_j N_j \phi_j \tag{9.25}$$

and

$$E_n = \sum_j N_j E_{n,j}, \tag{9.26}$$

where N_j is the interpolating function associated with node j, and (ϕ_j, $E_{n,j}$) are the nodal values. Equation 9.18 is then expressed as the sum of the integrals on the boundary Γ_e of element e:

$$c\phi(r) = -\sum_e \left[\int_{\Gamma e} w(\mathbf{r}, \mathbf{r}_s) \left(\sum_j N_j E_{n,j} \right) \mathrm{d}S - \int_{\Gamma e} \frac{\partial w(\mathbf{r}, \mathbf{r}_s)}{\partial n} \left(\sum_j N_j \phi_j \right) \mathrm{d}S \right]. \tag{9.27}$$

By using a numerical integrating scheme, such as Gauss-Legendre quadratures, we can write Eq. 9.26 in matrix form as

$$c\phi(\mathbf{r}) = \begin{bmatrix} h_1 & h_2 & \cdots & h_m \end{bmatrix} \begin{bmatrix} \phi_1 \\ \phi_2 \\ \vdots \\ \phi_m \end{bmatrix} + \begin{bmatrix} g_1 & g_2 & \cdots & g_m \end{bmatrix} \begin{bmatrix} E_{n,1} \\ E_{n,2} \\ \vdots \\ E_{n,m} \end{bmatrix}, \tag{9.28}$$

where m is the number of nodes on Γ.

Applying Eq. 9.28 to all nodes on Γ yields a system of linear equations,

$$\begin{bmatrix} c_1\phi_1 \\ c_2\phi_2 \\ \vdots \\ c_m\phi_m \end{bmatrix} = \begin{bmatrix} h_{1,1} & h_{1,2} & \cdots & h_{1,m} \\ h_{2,1} & h_{2,2} & & \\ \vdots & & \ddots & \vdots \\ h_{m,1} & & \cdots & h_{m,m} \end{bmatrix} \begin{bmatrix} \phi_1 \\ \phi_2 \\ \vdots \\ \phi_m \end{bmatrix} + \begin{bmatrix} g_{1,1} & g_{1,2} & \cdots & g_{1,m} \\ g_{2,1} & g_{2,2} & & \\ \vdots & & \ddots & \vdots \\ g_{m,1} & & \cdots & g_{m,m} \end{bmatrix} \begin{bmatrix} E_{n,1} \\ E_{n,2} \\ \vdots \\ E_{n,m} \end{bmatrix}, \tag{9.29}$$

or simply, in matrix form,

$$[H][\phi] = [G][E_n], \tag{9.30}$$

where $[H]$ is a matrix which includes the terms c_i. The following relationship is useful for determining the diagonal elements of $[H]$, particularly where the boundary is not smooth at node i, for example, at a corner point:

$$c_i + \sum_{j=1}^{m} h_{i,j} = 0. \tag{9.31}$$

The above equation holds true for any closed region. From Eqs. 9.27 and 9.28, we can conclude that $h_{i,j}$ and $g_{i,j}$ in Eq. 9.29 are in fact dependent on geometrical parameters but not on the boundary values ϕ_j and $E_{n,j}$. Hence, if we assume that the region Ω_A is enclosed by an equipotential contour, the field must be zero inside and Eq. 9.31 must be satisfied.

Now consider closed region Ω_A and its boundary $\Gamma = \Gamma_1 + \Gamma_2$. The potential is given on boundary Γ_1, while the normal component of the field is given on Γ_2, i.e., ϕ_i and $E_{n,i}$ are known variables on Γ_1 and Γ_2, respectively. Equation 9.30 is then sufficient for obtaining the unknowns, which are potential ϕ_i at the nodes on Γ_2 and normal electric field $E_{n,i}$ at the nodes on Γ_1.

9.4.2 Composite Dielectric Cases

The configuration shown in Fig. 9.9 is composed of two regions Ω_A and Ω_B. Γ_{AB} is the common boundary (dielectric interface), whereas Γ_A and Γ_B are the outer

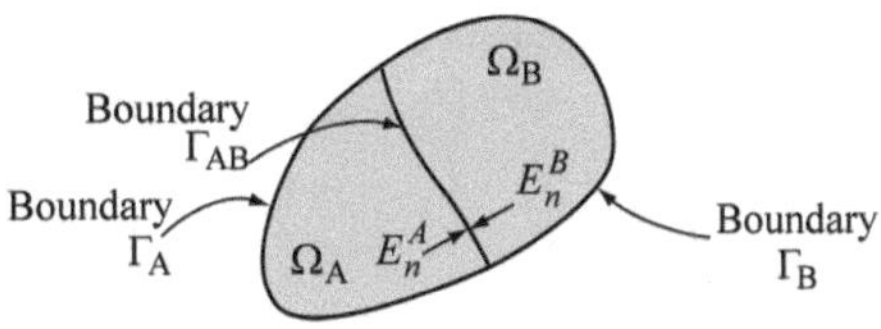

Fig. 9.9 Two regions Ω_A and Ω_B having common boundary Γ_{AB}

(uncommon) boundaries. Either the potential or the normal component of the electric field is given on Γ_A and Γ_B, while both the potential and the field are unknown on Γ_{AB}. The BEM solves the problem by applying (a) Eq. 9.30 to the nodes on each boundary, and (b) the following boundary conditions to the nodes on the common boundary Γ_{AB}:

$$\phi_i^A = \phi_i^B, \tag{9.32}$$

and

$$\varepsilon_A E_{n,i}^A + \varepsilon_B E_{n,i}^B = 0, \tag{9.33}$$

where ε is the dielectric constant. In these equations, A and B denote the side on which a value is taken. The normal electric field is positive in the direction outward from its corresponding region (see Fig. 9.9), which is opposite for Ω_A and Ω_B. This is the reason why Eq. 9.33 differs from the corresponding boundary condition given in the other sections of this book, e.g, Eqs. 1.10 and 9.4. Equations 9.32 and 9.33 give the additional set of linear equations that we need for solving all the unknown nodal values on Γ_A, Γ_B, and Γ_{AB}.

For a dielectric with volume conductivity σ, we replace the solid dielectric constant ε with a complex value,

$$\dot{\varepsilon} = \varepsilon + \frac{\sigma}{\mathrm{j}\omega\varepsilon_0}. \tag{9.34}$$

9.4.3 *Infinite Domain with an External Field*

In an open region with an infinitely remote boundary, the potential $\phi(\boldsymbol{r})$ is given as

$$c\phi(\mathbf{r}) = -\int_\Gamma w(\mathbf{r}, \mathbf{r}_s) E_n \mathrm{d}S - \int_\Gamma \frac{\partial w(\mathbf{r}, \mathbf{r}_s)}{\partial n} \phi \mathrm{d}S + \mathbf{E}_0 \cdot (\mathbf{r}_0 - \mathbf{r}), \tag{9.35}$$

where $\mathbf{r}_0$ is the reference point of zero potential. The equation is the same as Eq. 9.18, but it includes $\mathbf{E}_0$, an externally applied uniform electric field, when it exists in the region. Any other external field can be incorporated to Eq. 9.18 in a

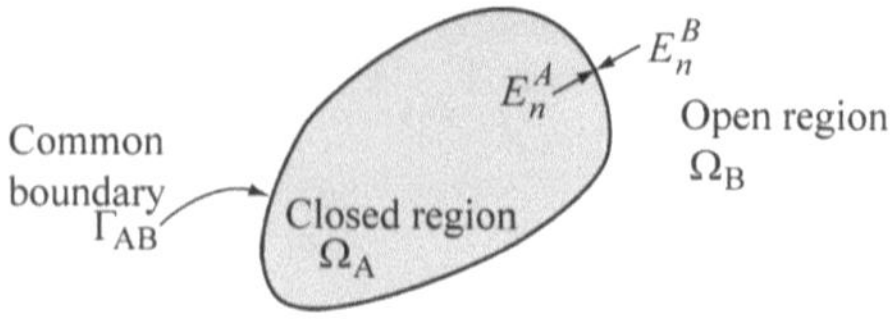

Fig. 9.10 A closed region Ω_A and an open region Ω_B

similar manner. It is to be noted that the term for an external field is not applied to closed regions. If the configuration is composed of closed region Ω_A and open region Ω_B, such as that shown in Fig. 9.10, a linear equation system can be constructed by applying

(a) Eq. 9.18 to nodes on the boundary of Ω_A
(b) Eq. 9.35 to nodes on the boundary of Ω_B, and
(c) Eqs. 9.32 and 9.33 to nodes on the common boundary Γ_{AB}.

In Fig. 9.10, however, all the nodes are on Γ_{AB}.

9.4.4 Dielectric Interface with Surface Conduction

For dielectric interface Γ_s with surface conductivity σ_s, the continuity of the potential in Eq. 9.32 still holds good, but the condition of the normal electric field must be modified to include the effect of the surface conduction. As explained in Section 1.3.2, the modified equation is

$$\dot{\varepsilon}_A \dot{E}_{nA} + \dot{\varepsilon}_B \dot{E}_{nB} = \nabla_s \cdot \dot{\mathbf{J}}_s = -\nabla_s \cdot \left(\sigma_s \nabla_s \dot{\phi} \right). \tag{9.36}$$

Directly applying this condition to the BEM requires evaluation of the tangential Laplacian term of $\dot{\phi}$, i.e., $\nabla_s \cdot (\sigma_s \nabla_s \dot{\phi})$, which may lower the calculation accuracy. As a countermeasure, the weighted residual method can be used with a set of weight functions g, resulting in the following equation [19]:

$$\int_{\Gamma s} g \left(\dot{\varepsilon}_A \dot{E}_{nA} + \dot{\varepsilon}_B \dot{E}_{nB} \right) dS = - \int_{\Gamma s} g \left(\nabla_s \cdot \left(\sigma_s \nabla_s \dot{\phi} \right) \right) dS. \tag{9.37}$$

Applying Green's theorem to the above equation yields

$$\int_{\Gamma s} g \left(\dot{\varepsilon}_A \dot{E}_{nA} + \dot{\varepsilon}_B \dot{E}_{nB} \right) dS = \int_{\Gamma s} \sigma_s \nabla_s g \cdot \nabla_s \dot{\phi} dS - \int_{ls} \sigma_s g \nabla_s \dot{\phi} \cdot \mathbf{n} dl, \tag{9.38}$$

where $\mathbf{n}$ is the normal vector on l_s, the boundary of Γ_s. In Eq. 9.38, only the tangential divergence operation is required in the BEM for handling problems including surface conduction. A convenient choice of the function g is the shape functions of the elements on the conductive surface.

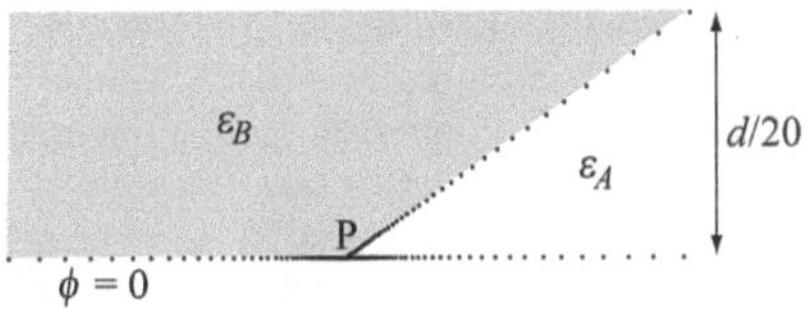

Fig. 9.11 Example of boundary subdivision near contact point P used for BEM calculation

9.4.5 Example of Boundary Division

In a similar approach to that of the charge simulation method (CSM), the boundary must be divided into denser elements as a contact point is approached so as to express the field singularity there. Figure 9.11 demonstrates an example of boundary subdivision used for computing the electric field where the contact angle is between 0 and $\pi/2$ in a 2D configuration. In the figure, media A and B, having a common boundary (line PQ in Fig. 2.6 or 9.6), are located between two parallel plane electrodes. The size of elements is represented by the distance between adjacent dots in the figure. The elements are concentrated near the contact point P by diminishing the element size in a geometric progression as the distance to the contact point decreases.

References

1. Takuma, T., Hamada, S.: Fundamentals and applications of numerical calculation methods of electric fields, Chap. 7. Tokyo Denki University Press (2006) (in Japanese), originally, Kouno, T., Takuma, T.: Numerical calculation method of electric field, Chap. 8. Corona Publ Co., Ltd. (1980) (in Japanese)
2. Takuma, T., Kawamoto, T.: Recent developments in electric field calculation. IEEE Trans. Magn. **33**(2), 1155–1160 (1997)
3. Takuma, T., Hamada, S., Techaumnat, B.: Recent developments in calculation of electric and magnetic fields related to high voltage engineering. Proc. 12th ISH (Int. Symp. on High Volt. Eng.) **1**, 1–11 (2001)
4. Steinbigler, H.: Anfangsfeldstärken und Ausnutzungsfaktoren rotationssymmetrischer Elektrodenanordnungen in Luft. Diss. Tech. Univ. Munich (1969) (in German), and Steinbigler, H. Digitale Berechnung elektrischer Felder. ETZ-A 90(25), 663–666 (1969) (in German)
5. Singer, H., Steinbigler, H., Weiss, P.: A charge simulation method for the calculation of high voltage fields. IEEE Trans. Power. Appar. Syst. **93**, 1660–1668 (1974)
6. Weiss, P.: Rotationssymmetrische Zweistoffdielektrika. Diss. Tech. Univ. Munich (1972) (in German)
7. Takuma, T., Konya, M., Hara, T., Kakimoto, N., Kawamoto, T.: Charge simulation method using fictitious charges at surrounding boundaries. Proc. 10th ISH **6**, 22–30 (1997)
8. Takuma, T., Kawamoto, T., Fujinami, H.: Charge simulation method with complex fictitious charges for calculating capacitive-resistive fields. IEEE Trans. Power. Appar. Syst. **100**(11), 4665–4672 (1981)
9. Takuma, T., Kawamoto, T.: Field calculation including surface resistance by charge simulation method. Proc. 3rd ISH:No.12.01 (1979)

10. Takuma T, Kouno., T., Matsuda, H.: Field behavior near singular points in composite dielectric arrangements. IEEE Trans. Electr. Insul. **13**(6), 426–435 (1978), and Kouno, T., Takuma, T.: Numerical calculation method of electric field, pp. 118–120. Corona Publ. Co., Ltd. (1980) (in Japanese)
11. Kawamoto, T., Takuma, T., Goshima, H., Shinkai, H., Fujinami, H.: Triple-junction effect and its electric field relaxation in three dielectrics. IEEJ (Inst. Electr. Eng. Jpn.) Trans. FM **127** (2), 59–64 (2007) (in Japanese)
12. Gutfleisch, F., Singer, H., Förger, K., Gomollon, J.A.: Calculation of high-voltage fields by means of the boundary element method. IEEE Trans. Power Deliv. **9**(2), 743–749 (1994), and Vetter, C., Singer, H.: High-voltage field computation using bi-cubic surface splines. Proc. 11th ISH **2**, 79–82 (1999)
13. Farin, G.: Curves and surfaces for computer aided geometric design, 4th edition, pp. 279–316. Academic Press, CA (1997)
14. Hamada, S., Takuma, T.: Application of a rational quartic triangular Bézier patch to field calculation by a boundary subdivision method. Int. J. Appl. Electromagn. Mech. IOS Press **13**, 431–435 (2001)
15. Hamada, S., Takuma, T.: Surface charge method using a triangular Bézier patch with a variable interior control point. Proc. COMPUMAG (Conf. on the Computation of Electromagnetic Fields), Lyon, PA3-7:I-82-83 (2001)
16. Tatematsu, A., Hamada, S., Takuma, T.: Analytic expressions of potential and electric field generated by a triangular surface charge with a high-order charge density distribution. IEEJ Trans. FM **121**(4), 378–384 (2001) (in Japanese)
17. Hamada, S., Takuma, T.: Electric field calculation in composite dielectrics by surface charge method based on electric flux continuity condition. Electrc. Eng. Jpn. **138**(4), 10–17 (2001)
18. Brebbia, C.A.: The boundary element method for engineers, 2nd edition. Pentech, London (1984)
19. Nicolas, A., Rasolonjanahary, I., Krahenbuhl, L.: Computation of electric fields and potential on polluted insulators using a boundary element method. IEEE Trans. Magn. **28**(2), 1473–1476 (1992)

Index

FSC
www.fsc.org
MIX
Papier | Fördert gute Waldnutzung
FSC® C083411

Kolibri 360 GmbH
Ferdinand-Jühlke-Straße 7
99095 Erfurt, Deutschland
produktsicherheit@kolibri360.de